Secrets of the Seas

ILLUSTRATED GUIDE TO
MARINE LIFE OFF SOUTHERN AFRICA

Compiled by
A.I.L. PAYNE & R.J.M. CRAWFORD

Illustrated by
A.P. VAN DALSEN

VLAEBERG PUBLISHERS

Contributors to colour illustrations

Numbers are given in two parts. In each instance the first part refers to page number and the second to the number of the photograph on the page

Adams, N. J. 71.3 71.5 71.6 74.4 74.6 74.7 75.4 75.9
Agenbag, J. J. 11.1 11.2 11.3
Armstrong, M. J. 26.3 27.2
Augustyn, C. J. 19.3 22.2 22.3 22.4

Bannister, A. 51.2 95.4 95.5 95.6
Berruti, A. 10.4 55.4 59.7 62.2 62.3 74.5 102.6
Best, P. B. 83.6 86.4 86.5 87.2 90.2 90.4 90.5 91.6 94.1 94.3 94.4
Brokensha, M. M. 87.8
Buxton, C. D. 38.6 50.3

Compagno, L. J. V. 22.5 39.4 43.3
Crawford, R. J. M. 10.5 15.3 54.5 55.1 55.2 55.3 55.5 55.6 58.4 58.6 58.7 59.1 59.2 62.1 62.5 63.1 63.4 63.6 66.3 66.6 66.7 66.8 67.7 67.8 78.1 78.4 99.6 102.5
Cruickshank, R. A. 14.3 30.3 66.5 75.1 75.2 87.4 90.3

David, J. H. M. 7.4 78.5 79.1 79.2 82.1 82.2 82.3 82.5
Davis, B. 47.4
Dawbin, W. H. 98.1
Dyer, B. M. 26.6 58.3 58.5 59.4 62.4 62.7 63.5 67.1 67.5 79.3 83.4 86.3 87.1 99.3 99.4 102.2 102.4 Front Cover (Heaviside's dolphin, South African fur seal, African penguin), Back Cover (swift tern)

Fridjhon, G. H. 83,5

Garratt, P. A. 42.1 50.2
Gates, J. 30.1 70.5 71.1 71.2 75.6 75.7 86.6 86.7 87.3 99.2
Girzda, J. 42.4
Goosen, P. C. Front Cover (knifejaw)
Granger, E. 10.1 10.2 10.3
Grant, W. S. 7.1 30.2 66.1 102.1

Harding, R. T. 19.6 23.4 23.6 35.2
Heydorn, A. E. F. 51.3 54.2
Horstman, D. A. 14.1 14.4 14.5 15.1 15.2 15.6 18.5
Hughes, G. R. 51.1 51.4 54.1 54.3 54.4 95.7

Japp, D. W. 19.5 30.6 31.1 31.2 31.3
Joubert, P. A. 22.6 38.1 38.2 38.3 38.4 39.5 42.3 95.3 98.2 98.3 98.4

Leslie, R. W. 30.4 43.4 67.2 75.8 75.10

Malan, P. E. 7.2 59.5 59.6 70.1 70.2 70.3 70.4 70.6 70.7 74.1 74.3 74.8 75.3 75.5 79.4
McCartney, C. 91.2
Meÿer, M. A. 7.5 71.4 74.2 78.3 78.6 79.5 82.4 82.6 82.7 83.3 87.6 87.7 90.6 91.1 91.3 91.4 91.5 94.2 94.5 94.6 99.5
Miros, P. (courtesy South African Navy) 46.1 46.2 46.3 46.4 46.5
Moldan, A. G. S. 18.2 18.4 19.1

Natal Mercury 94.7
Natal Parks Board 46.6
Natal Sharks Board 27.4 27.5 27.6

Oceanographic Research Institute 46.7 47.1 47.2 47.3 50.1 51.5

Paterson, J. 99.7
Penney, A. J. 27.3 31.4 34.1 34.2 34.3 34.6 50.6 63.7 63.8 66.2 66.4 67.6 98.5 98.7
Pitman, R. 87.5
Pollock, D. E. 23.1 23.2 23.3 23.5

Quartin, N. 35.5

Randall, J. 42.2
Randall, R. M. 18.3 58.1 58.2 59.3 62.6 67.3 67.4
Rose, B. 7.6
Ross, G. J. B. 86.1 90.1
Rowe, S. 14.2

Schülein, F. H. 7.3 15.5 19.2 19.4 26.2 26.4 26.5 27.1 78.2 82.8 99.1
Smale, M. J. 35.6 38.5 39.1 39.3 42.5 43.1 43.2 43.5 50.4
Smith, C. 35.4
Sutherland, B. 63.3 83.1
Swart, R. 86.2

Tarr, R. J. Q. 15.4 22.1 34.4 34.5 35.1 98.6
The Argus 18.1
Thurman, G. 39.2

Van Dalsen, A. P. 26.1 30.5
Van der Elst, R. P. 31.5 42.6 50.5 102.3
Visser, B. 63.2
Visser, J. 79.6

Wahl, Z. 83.2 95.1 95.2
Williams, P. V. G. 35.3

Six animals are illustrated on the front and back covers. On the front cover are four endemic species, a juvenile knifejaw, a South African fur seal, two African penguins and the only species of dolphin found nowhere else than off Southern Africa, the Heaviside's dolphin. On the back cover is a photo of swift terns, of which our subspecies breeds only in the subcontinent, and a drawing of a great white shark. All photos except one were taken by Bruce Dyer, the exception being the photo of the knifejaw, which was taken by Pedro Goosen.

CONTENTS

Acknowledgements 5
Introduction to the resources of the open sea 6
Estuaries 8
The physical environment 9
Phytoplankton, the small plants of the sea 12
Zooplankton, the small animals of the sea 13
Marine pollution 16
The major fisheries 17
Molluscs 20
Spiny lobsters and prawns 21
Purse-seine fishing 24
Purse-seined shoaling fish 25
Bottom trawling 28
Longlining 29
Coastal fishing 32
Linefish 33
Bony fish 36
Sharks and rays 41
Subsistence and recreational fishing 48
Sea turtles 49
The coastal islands and guano platforms 53
African penguins 56
Gannets, cormorants and pelicans 60
Gulls and terns 65
Migrant seabirds 72
Seals 76
Cetaceans 81
Marine reserves 96
Marine ecosystems 100
The heritage 101
Index 103

To our parents, especially those who are not with us to see the finalization of this work. Through their early provision, they are in large measure responsible for it.

ACKNOWLEDGEMENTS

Cape gannet

This book was conceived as a follow-up to the far more comprehensive *Oceans of Life off Southern Africa*. It is intended as a more affordable option to its predecessor, especially for the young readers, although we hope it will be of interest to everyone.

The colour illustrations are the same as in *Oceans of Life off Southern Africa* to keep the price at a reasonable level. We owe our gratitude to the many photographers who supplied us with the excellent colour material reproduced here, and who gave permission to use it again.

The text is new. However, it could not have been written without knowledge gained from the various chapters of "Oceans", and we are also indebted to the many authors of that book who willingly gave their permission to use their work as a basis for this text.

Others deserve mention as well: we were consistently supported by our families, the Board of Trustees of the Offshore Resources Book Fund and our seniors and colleagues at the Chief Directorate: Sea Fisheries and particularly in the Sea Fisheries Research Institute.

Three of our colleagues deserve mention by name, Hajo Boonstra, Bruce Dyer and Irma van der Vyver. They assisted us with the editing and proof-reading of the text and in collating the artwork. The foregoing and our publisher, Mr G. de Melker, saw to it that this book reached fruition, and they deserve our sincerest thanks.

INTRODUCTION TO THE RESOURCES OF THE OPEN SEA

A crowned cormorant nesting

Unlike many of the seas of the northern hemisphere, the waters around southern Africa have only really been known well for about 500 years. The Portuguese arrived first and were followed by the Dutch and British. These Europeans soon started to exploit the animals of the region, but few then understood that such resources are not inexhaustible, in other words, that there is a limit to how many animals can be caught.

Although the initial level of utilization was low, news of the bountiful resources soon leaked back to Europe and America, fleets were despatched and overutilization quickly followed. Seals, whales, seabirds and, more recently, fish and shellfish have been overutilized, and now some resources are no longer economically viable.

Early legislation was unsuccessful in protecting them, largely because the seas are so unpredictable and man is so greedy. Even now, much remains unknown about the sea, but most people believe in controlled utilization rather than straight exploitation, and that gives us some optimism for the future of our resources. Here, we have called them our resources, but the word "our" is a misnomer. No-one owns marine resources outright, so careful utilization is essential to maintaining them for the use or enjoyment of future generations.

What then are the marine resources of southern Africa? It would take more than this book to answer the question, such is the diversity of marine life around our shores, but here we will try to introduce you to the more prominent resources. The photographs opposite give three examples. The fur seal was among those species severely depleted in the past, as were some whales. Whales continue to be relatively scarce off our coast, despite there being an almost total moratorium on their capture worldwide.

By contrast, the South African fur seal provides an example of how a resilient resource can bounce back after devastating overutilization. Having been hunted to the point of economic non-viability, there are now over a million seals living around our shores. Their competition with man for some of the valuable fish populations often causes impassioned pleas from fishermen for their control. Complete extermination would never be the answer, because fur seals are an integral part of the marine system around our coast and their extinction would not necessarily benefit the fisherman.

Crowned cormorants have no direct economic value, but are of recreational and conservational interest. Southern Africa possesses a wealth of wonderful, occasionally very visible, marine animals, some of which, for example the crowned cormorant, occur nowhere else. Many of the less numerous species can be affected by man's depredations on the abundant resources.

The map on this page shows the main areas of fishing for some of southern Africa's important marine resources. Only some species of fish can be caught from the shore, but many more are caught by boat, using various forms of fishing gear.

We hope that, after reading about the various resources, you will begin to understand just how important it is to conserve them for future generations.

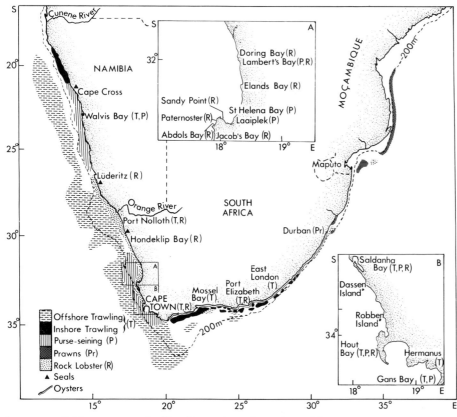

Important fishing areas and offloading points of southern Africa

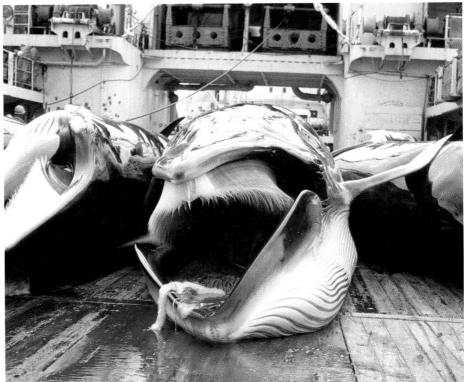

Some of the resources of the open ocean may be harvested from land, e.g. by anglers **1**, but craft of some sort are usually necessary. They may be small, as are rock-lobster dinghies **2**, or large, as are distant-water trawlers and their supply vessels **3**. Among the earliest species to be intensively exploited were the more visible South African fur seals **4** and whales **5** - minke whales aboard a factory ship. Many of the animals of southern Africa's seas are endemic, e.g. crowned cormorants **6**, one of the world's rarest cormorants

ESTUARIES

Shixini River estuary

Estuaries are bodies of water with intermittent or permanent connections to the sea. They are dealt with here because much of man's utilization and enjoyment of the sea takes place in and around estuaries and because his influence on them can be great.

In South Africa, there are 343 estuaries, most of which are along the East Coast. In fact, two-thirds of South Africa's total estuarine area of some 600 km^2 is in Natal. Many of these East Coast estuaries are subject to flooding, whereas many West Coast estuaries are cut off periodically from the sea or sometimes even dry up. In Namibia, the pans at the coast are the dried-up estuaries of rivers.

Obviously, the rivers draining to the East Coast, because of their habit of flooding and because they are steeper and flow faster, tend to carry more terrestrial sediments to the sea; therefore, their influence on the coastal marine resources is likely to be greater. The Orange River, because it has its source in the east of the country, also carries a large load of terrestrial sediments to the sea.

Many marine animals are dependent on estuaries at some stage of their life. Birds often nest in estuaries, and some fish and shellfish periodically migrate between the sea and freshwater to breed, to feed or for protection. Prawns are much sought-after as food, but they need estuaries to survive, as shown in the drawing. Eggs are laid at sea, but even before it has assumed its adult shape, the young prawn has migrated into the estuaries to feed and grow. Only when it has grown does it return to the sea to breed.

In addition to prawns, 81 of South Africa's fish species depend wholly or partially on estuaries for their existence. Of these 81, 29 are sport-angling fish and another 21 are utilizable as food. As with the prawns, their reproductive cycles are all synchronized with the coastal climate, which determines seasonal opening and closing of the estuary mouths.

Many West Coast estuaries are important feeding grounds for some birds, which migrate south from the northern hemisphere to escape the cold northern winters.

Many people are drawn to estuaries for tourism, and the infrastructure dependent on tourism is itself a large industry. Would these same holidaymakers be so keen to visit if the fish were not so abundant, or if the estuaries were polluted?

By the mid 1980s, only 24% of Cape estuaries and 28% of those in Natal were still in a good condition. In contrast, 20% of the Cape estuaries and 26% of those in Natal were already classified as being in a poor state. All the ill-effects, for example pollution, careless road and rail building, inadequate water regulation in catchment areas, badly planned residential and industrial development in key areas and poor agricultural practices, were caused by man.

Many of southern Africa's estuaries are attractive places. However, to ensure that tourists still visit them and to protect the marine resources which depend on them, it is vital that environmental conservation along the coast (including the estuaries) and marine resource management go hand in hand. To date, this has not been the case.

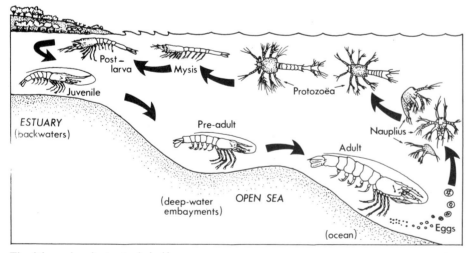

The life cycle of a typical shelf prawn

THE PHYSICAL ENVIRONMENT

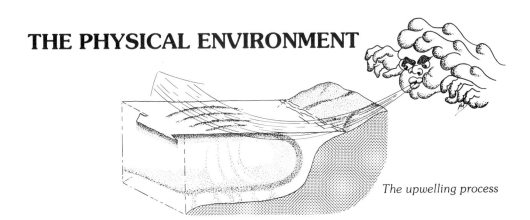

The upwelling process

We are privileged in southern Africa to have one of the most varied marine environments in the world next to us. On the East Coast, the warm Moçambique and Agulhas currents flow south-westwards and, on the West Coast, cold surface water moves north (the Benguela system). These currents, the effect of which can be seen clearly in the satellite images on p. 11, influence the climate of southern Africa and contribute to the weather along the West and East coasts being so different. The physical oceanographic environment is highly variable, often causing the abundance of some marine resources to fluctuate quite dramatically over comparatively short periods.

The Agulhas Current is part of the large-scale circulation of water in the southern Indian Ocean that is driven by solar energy, through different rates of heating and cooling at the equator and at the poles, and the resultant large-scale winds. It moves at an average of 1-2 metres per second and transports some 60 million cubic metres of water every second, a quarter of a million times the average flow of the Orange River. Generally it extends as far as the western edge of the Agulhas Bank (see the large satellite image on p. 11), where it doubles back on itself to flow eastwards farther south in what is known as the Agulhas Return Current.

Sometimes, though, its warm water rounds the Cape, enters the Atlantic, and moves north. It needs little imagination to predict changes in availability of fish stocks on the West Coast (where the water is mainly cool near shore) when warm water arrives from the South Coast. Distributions and population sizes of particularly short-lived species, such as anchovy, can be drastically altered.

The Benguela is not a single deep current, a "river in the sea", like the Agulhas, but rather a complex system driven by trade winds towards the equator. It is characteristically cool at the surface and very productive of biological material. That, of course, is why the Benguela and similar systems are centres of some of the biggest fisheries in the world. The reason for the highly productive nature of the Benguela system is shown conceptually in the drawing on this page. The driving force is the south-east wind, which those who live in the Cape or along the West Coast probably detest when it blows strongly for days on end. Without it, however, there would be little production of the small plants and animals (the plankton) on which most fish rely for food, either directly or indirectly.

Upwelling, the process which brings nutrients (fertilizers for the plants) to the surface, occurs where deep water is close to the coast and where strong longshore winds initiate offshore movement of water. Lüderitz is the strongest site of upwelling, but there are other centres off Namibia and South Africa, the latter including Cape Columbine (Saldanha) and the Cape Peninsula. Once nutrients reach the upper, sunlit parts of the ocean they can be used by plants to produce food in abundance for herbivorous animals. This process also involves carbon dioxide and is known as photosynthesis. The average speed of flow of the surface Benguela water is about 15-20 centimetres per second. It propels about 15 million cubic metres of water per second towards the equator, only a quarter of the volume of the Agulhas Current, but the widescale upwelling makes it extremely important for southern Africa's marine resources.

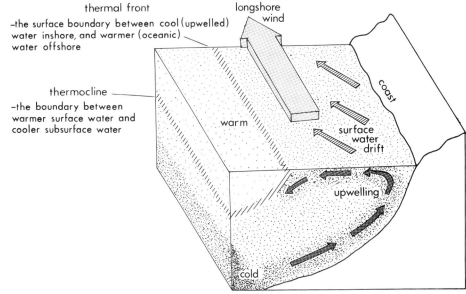

Schematic illustration of upwelling

Most estuaries adjoining southern Africa's oceans are attractive, e.g. the Kobole **1**, the Mbashe **2** and the Mpako showing Hole-in-the-Wall **3**. Some seabirds breed in estuarine systems, including white pelicans at Lake St Lucia **4** and crowned cormorants at Schaapen Island in the Langebaan Lagoon **5**, here nesting with blackheaded herons, cattle and little egrets

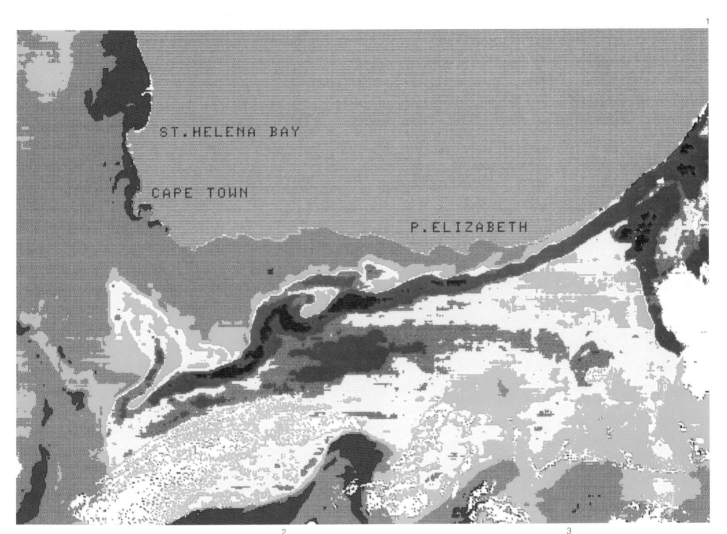

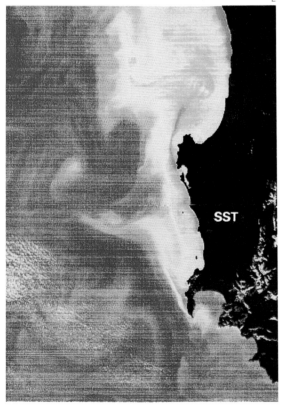

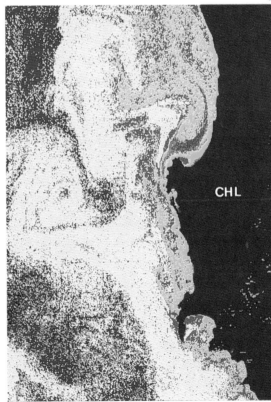

Satellite images of sea surface temperature around South Africa on 16 July 1979 **1**, and of sea surface temperature **2** and chlorophyll **3** off the Western Cape, on 9 May 1979. Evident on the first are the warm Agulhas Current (red) on the South Coast and the cold upwelled water of the Benguela (dark blue) on the West Coast. The bottom images show a mosaic typical of the highly productive Benguela system

PHYTOPLANKTON
The small plants of the sea

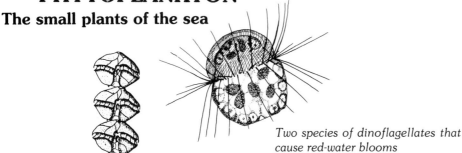

Two species of dinoflagellates that cause red-water blooms

Plankton is the name given to plants (phytoplankton) and animals (zooplankton) whose powers of locomotion are too weak to prevent them from being transported by currents. Many phytoplankton are very small, perhaps 0,02-0,2 mm long, but some species unite into chains which make them bigger and more visible to the naked eye.

Because plants are the only form of life capable of photosynthesis, in other words capturing energy from the sun for converting nutrients into living matter, they form the first trophic (food) level in the ocean. All the other levels depend on them – the drawing on this page shows the interrelationships of the various trophic levels.

Some of the phytoplankton species are spectacularly shaped but have to be viewed under a high-powered microscope. One group, the diatoms, are particularly successful in growing rapidly in nutrient-enriched water, and it is their greeny pigments which alter the colour of the sea to a murky green-brown as they "bloom". Photosynthesis in the sea occurs only near the surface where there is sufficient light. Upwelling, mixing (for example by storms bringing subsurface waters nearer the surface), and rivers flowing into the sea enhance the supply of nutrients, thereby allowing phytoplankton to bloom. Nutrients are then rapidly used up and the murky bloom reduces the penetration of light. The result is that the production of plant material soon slows down – unless the winds start to blow again, which is exactly what happens on the windy West Coast in spring and summer, those seasons when productivity is at its maximum. To most of us, such plant blooms in the sea mean little. However, it has been estimated that all the oceans of the world together produce a billion tons of plant tissue per year, most of this in the 10% of the ocean surface adjacent to land masses. Fortunately for the marine resources which depend on this production, none of it can yet be harvested economically.

One phenomenon which may affect man and which is caused by phytoplankton blooms is the so-called red water, which sometimes follows periods of calm, settled weather. Examples are shown on p. 14. The dinoflagellates (another group of phytoplankton), which have green, yellow, orange, red or brown coloration, are usually responsible for this red water. Certain of them have a toxic (poisonous) substance in their cells, and these cells, if filtered out of the water by filter-feeders such as mussels, can have several direct or indirect effects on man and the nearshore and intertidal fauna. Where such mussels are then consumed, the effects can be manifest as paralytic, diarrhetic, neurotoxic or amnesic shellfish poisoning. Even cooking will not necessarily make the shellfish safe to eat, so the safest course is to leave such shellfish alone during summer or autumn (the main seasons of red water) and to heed the warnings issued in the press if a toxic dinoflagellate bloom is suspected. Many outbreaks of red water are non-toxic, and sometimes fish or shellfish die simply by suffocation as the phytoplankton cells, which are present in vast numbers, either clog their gills or use up most of the oxygen in the water. Also, some of the toxins do not necessarily affect the initial carrier (for example mussels) but can have serious effects on the final consumer. Fortunately, wind rapidly causes the dispersal of red water, in time making it safe once again to eat filter-feeding shellfish.

Food chains in the sea

Zooplankton are those aquatic animals which are unable to resist transport by currents. They feed either on the phytoplankton or on other small animals. Zooplankton organisms are not necessarily small, ranging in size from the 0,02 mm of some microscopic protozoans, through the swimming shrimps such as krill (on which whales feed), which can be several centimetres long, to large jellyfish. Also included in the zooplankton are the young forms of shellfish and the eggs and fry of fish, all of which tend to be at the mercy of currents and the weather. That, of course, is why so many shellfish and fish spawn numerous eggs – nearly all of the eggs and fry are either eaten, starve or are simply swept away to regions where the environment is not suitable for their survival.

Some zooplankton organisms which live around our coasts are shown on p. 15. The most primitive forms are the tiny one-celled protozoans, whose shells can accumulate over many years to form such spectacular deposits as the white cliffs of Dover in England. Then there are the jellyfish, of which the Portuguese man-of-war (or blue-bottle) is an example. It is actually a colony of specialized individuals, some for stinging, others to digest food, and others still to aid in flotation. The blue-bottle is therefore an amazingly complex "city in the sea", but beware, the sting is painful, even after the jellyfish has been on the shore for some time.

Also in the zooplankton are worms, the so-called glass worms or arrow worms (chaetognaths), which can consume fish fry their own size, and swimming molluscs such as the sea butterfly, distant relatives of the mussels you find along the shore. However, by far the most diverse and numerous group in the zooplankton is the arthropods (which means jointed legs). This group includes the shelled crustaceans, many forms of which live in our waters – six are shown in the drawings on this page. The copepods (literally translated "oarfeet") are by far the most abundant, contributing 60-80% of the total mass of zooplankton in the ocean. They eat the phytoplankton and are in turn eaten by bigger zooplankton and, sometimes, by fish. The next most important group is the euphausiids (meaning "shining light" because they glow in the dark) which, as can be seen from the plate on p. 15, can be found in vast swarms. Finally, the zooplankton contains some jelly-like animals which, because they have a rudimentary backbone, can be classified as primitive vertebrates. Some of these tunicates, as one group is known, could be mistaken for jellyfish but for the bands of muscle surrounding their barrel-shaped bodies.

The zooplankton have some interesting patterns of behaviour and, although there are many theories to explain these patterns, knowledge is still limited as to why they act in the way they do. Why, for instance, do they migrate up and down in the water column perhaps daily (to the surface at night and to depths by day) but sometimes not at all? Many fish do the same, so the migrations do not necessarily allow the zooplankton to avoid predators. Also, how do they move up and down so easily? Again, why do some have such powerful light-producing organs? Do these attract other members of the same species, act as a lure to prey, or confuse an attacker? Whatever may be the answers to these questions, zooplankton is clearly a very successful assemblage of animals.

Man should think carefully before he decides to harvest zooplankton, even in small quantities, because most large organisms in the ocean ultimately depend on it and we dare not radically alter the course of nature.

ZOOPLANKTON
The small animals of the sea

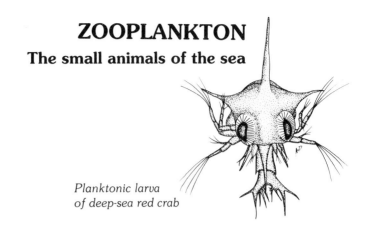

Planktonic larva of deep-sea red crab

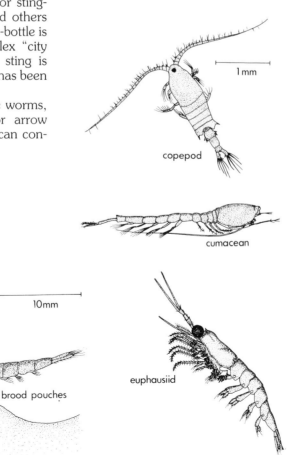

Red-water outbreaks are of interest because eating filter-feeding molluscs that have fed on organisms responsible for the red water may result in paralytic shellfish poisoning. A bloom of *Noctiluca miliaris* in False Bay taken with a polarizing lens is shown in **1**, and blooms of the same species taken from an aircraft and from a ship in **2** and **3**. A healthy bloom of the red-water species *Mesodinium rubrum* is not red **4**, whereas when it decays it is **5**

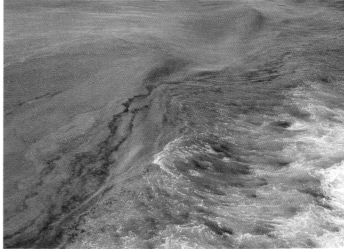

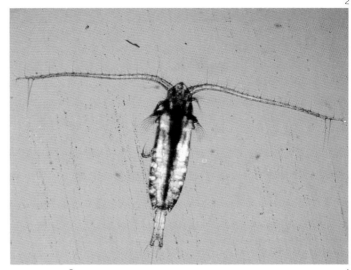

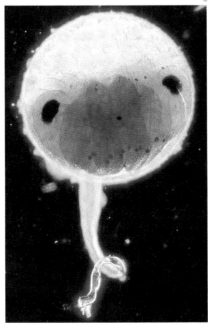

Euphausiids **1** and copepods **2** are among the more important food sources in the Benguela system, being fed on by many of the abundant fish species. Part of a huge swarm of euphausiids was stranded at Possession Island in December 1985 **3**, being packed along the beach in dense aggregations for almost a kilometre and providing a feast for various gulls and terns. Among the larger free-swimming invertebrates off southern Africa are the mantis shrimp **4**, sometimes an important prey of snoek and the Cape hakes, and jellyfish **5**, of ecological importance off the Namib coast. Some organisms are only planktonic for part of their life cycle, e.g. crab larvae **6**

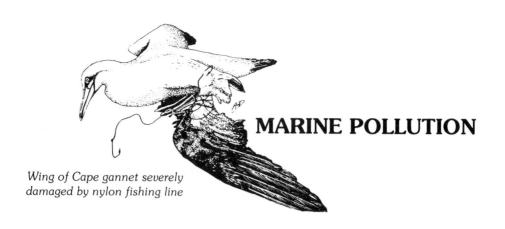

Wing of Cape gannet severely damaged by nylon fishing line

MARINE POLLUTION

One of the most destructive of man's habits is to pollute his environment. It is hard to believe that it was not until the 1950s that man first realized that to release waste materials in an uncontrolled manner into an aquatic environment often spelt disaster. Environmental contamination can, of course, be natural, for example the seepage of natural oil from the sea bed. Here, though, we consider pollution of the seas or estuaries by man. This can be described as the direct or indirect introduction of substances or energy that results in such deleterious effects as harm to living resources, hazards to human health, hindrance to marine activities including fishing, impairment of quality for use of seawater and reduction of amenities.

Pollution of the marine environment comes from sources on land and at sea. Domestic sewage, industrial effluents, and rivers contaminated by pesticides and fertilizers or containing high loads of silt caused, for example, by poor agricultural practices may all be discharged to sea. Ships at sea may deliberately or accidentally dump wastes, and oil spills from damaged tankers may cause heavy pollution. Exploitation of the mineral resources that occur on the sea bed may also cause pollution, and so too may precipitation from an atmosphere contaminated with gases and particulates. The hydrocarbons that originate from land-based sources, such as oil refineries, storm-water drains and car exhausts, probably have more serious effects on the sea than the occasional spectacular oil spill, but it must be remembered that South Africa is situated on one of the world's busiest shipping routes - 20% of Middle East oil exports at present pass the Cape of Good Hope. It is a wonder that we have not experienced more oil spills, especially in the 1970s when, with the Suez Canal closed and little exploitation of oil elsewhere in the world, about five times as much oil was rounding the Cape than at present. The drawing on this page shows how the quantity of oil transported around southern Africa has changed over the past few years.

The influence of pollution on marine life depends on the chemical or physical characteristics of the pollutant. How long it persists, whether it is degradable, the extent to which it accumulates in biological tissues and sediments, its solubility in water, the likelihood of its being transformed into even more harmful compounds, its effect on the quantity of oxygen in the water, and its toxicity are all important considerations. Many pollutants that have no effect over a short period can be most harmful if an animal is exposed to them for a long time. For example, accumulation of DDT in birds of prey has caused them to lay eggs with thinner shells which crack during incubation and therefore do not hatch. A different effect is caused by the effluent from fish factories, which reduces the quantity of oxygen in the water, sometimes leading to suffocation of certain marine animals.

The best-known toxic pollutants of the sea are metals, particularly mercury and lead, biocides (the pesticides, herbicides and fungicides) such as DDT, the insulating fluids in electrical and heat-transfer equipment (commonly known as PCBs - polychlorinated biphenyls), hydrocarbons such as those derived from oil, and radioactive compounds.

The coast of southern Africa is fortunate in being less industrialized than most coastlines of the northern hemisphere. It is also more exposed and there is deep water close inshore to help disperse and hence dilute the dangerous substances. Therefore, pollution does not, as yet, pose a major threat to our shores or, in most cases, to our marine life. Nevertheless, our authorities are not being complacent about the potential of the problem and all forms of pollution are being closely monitored, including the more visible aspects of pollution, for instance the oil and the plastics on the beaches. Control of the source of pollution is the only answer, and it is heartening that a very strong lobby is making itself heard worldwide and influencing politicians and industrialists to act against this potential disaster to mankind.

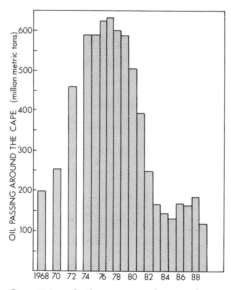

Quantities of oil transported around southern Africa, 1968-1989 (except 1969, 1971 and 1973)

THE MAJOR FISHERIES

Fishing for spiny lobster

Now that we have introduced you to some factors that influence the marine resources of southern Africa, it is time to introduce the major fisheries of the region. These, after all, contribute most of the catch being made around our coast.

There are many types of fishing gear, but three in particular catch large quantities of fish. These are the purse-seines, which catch small shoaling species such as pilchard (sardine) and anchovy; the bottom trawls, aimed at the fish you usually see displayed on a fish counter, for instance hake and sole; and the midwater trawls, in which most of the horse mackerel (or as it is often called, the maasbanker) are caught.

Fishing our rich seas, South Africa and Namibia combined have risen as high as sixth in the world's list of top fishing nations. Currently, with the catches of the purse-seiners being well below that of the halcyon days of the late 1960s, we are normally ranked about twentieth. Nevertheless, we are still a significant fishing power. Until 1977 in South Africa, many other countries also fished our waters, mainly for hake and horse mackerel. Then, of the total catch of these species, our local fleets only took about one-third. Not all the local harvest is eaten – much goes to fishmeal – and South Africans still are not traditionally big fish-eaters.

The colour plate on p. 19 illustrates working conditions on fishing boats, and shows some of the animals that are caught. It is of interest to compare the importance of the various fisheries. In terms of quantity caught the purse-seine fishery comes first, but in terms of value, the demersal (bottom trawl) catch far outstrips it. Catches of spiny lobster, prawns caught by trawl and squid (listed as linefish) are small, but their value is high. Little wonder then that there is a permanent waiting list of potential fishermen seeking a permit to try for their slice of the profit. The total mass caught by the fisheries in South Africa in 1989 was 665 504 tons, and its value at first sale was R1 104 million. With such a large industry, there are entire communities and even towns that depend almost totally on revenue derived from fishing.

The decision-makers in southern Africa have fortunately realized that the valuable fish resources are not infinite, and catches are carefully controlled. That is why limited numbers of permits are granted for the various fisheries. Permits, of course, are only one of the means of control applied. Catches may be governed by means of quotas, meshes in a net may have to be larger than a specified minimum size (to allow young fish to escape), and sanctuaries where no fishing is allowed may be demarcated. There may be a closed season, a minimum size specification requiring undersize animals to be returned to the sea, or even a limitation on the type of gear that can be used. For instance, no-one is allowed to use SCUBA to catch spiny lobsters. Finally, to protect the consumer by ensuring good condition of fish, it is necessary sometimes to set condition limits on the fish. Thus, spiny lobsters in berry may not be landed.

All these restrictions help to ensure that the waters of southern Africa remain among the top fishing grounds in the world.

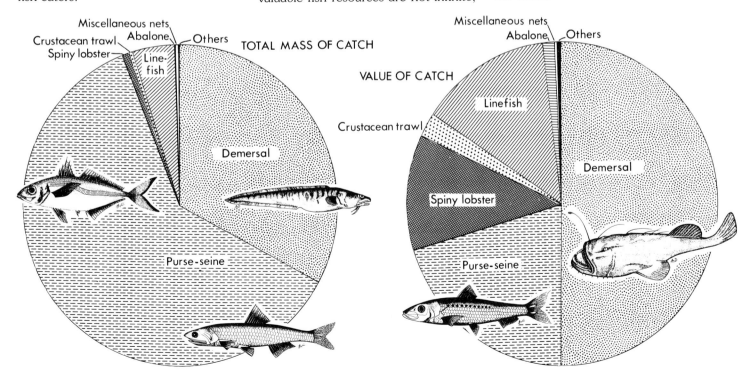

Relative contribution by different sectors to the mass and value of South Africa's fish catch in 1989

In 1983 the oil tanker *Castillo de Bellver* caught fire off Saldanha Bay **1**. Oil spilt in such accidents is often cast up on beaches to the consternation of local authorities **2**, but it is the animals that may suffer most, e.g. an African penguin **3**. Effluent from fish factories is another source of pollution **4-5**

The more important fisheries off southern Africa include purse-seining for species such as anchovy and pilchard that shoal near the surface **1**, midwater trawling for horse mackerel mainly by foreign fleets **2**, bottom trawling for hakes **3**, longlining for tuna **4** (a bigeye tuna being brought aboard with a gaff) and kingklip **5** (preparing the lines ready to be set), and inshore fishing for species such as spiny lobster and hottentot **6**. Longlining for kingklip has been phased out

MOLLUSCS

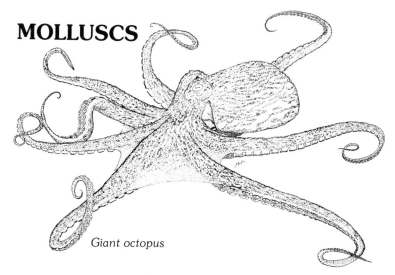

Giant octopus

The well-known garden snail is an example of a mollusc. There are many molluscs in the sea and on our beaches. Mussels, oysters, periwinkles and limpets you can find on any visit to the seashore, but none of them support large fisheries. That description does, however, befit two of our molluscs, the abalone and the chokka squid, one of many species of squid found in our waters. These two animals are shown on p. 22, together with a living paper nautilus (have you ever found the beautiful shell of this animal on the beach?) and a cuttlefish, an animal which, like the octopus, is related to squid.

Abalone, or perlemoen as most South Africans know it, are probably best known locally for the unusual flat shell with its mother-of-pearl lining, a shell used for such diverse purposes as soap dishes, plant containers, ashtrays and food holders. The flesh is, however, popular both in the Far East and locally. In southern Africa, strandlopers were taking them for food long before the present-day fishery was established. They left piles of cleaned shells to prove it. Nowadays the commercial fishery is permitted to take about 600 tons per year, mostly along the coastline near Hermanus, where the animal is abundant. Divers operating from dinghies take most of the commercial catch, but they are very dependent on the weather, which is suitable for their operations only about five days per month.

Abalone have an interesting life cycle. They spawn by releasing huge quantities of eggs and sperm into the water. All abalone in a particular area do this simultaneously. Once hatched, the larvae drift in the water for a few days. If lucky, they settle on a suitable area of the sea bed, where they turn into miniature adults. To feed, they trap pieces of drifting seaweed under their feet and graze on it. Young abalone have to move rapidly into nooks and crannies to avoid being eaten, but large ones rarely move far because the currents bring food to them. Carefully marked large abalone have, in fact, been found in the same position on the same rock for several years.

Squid, in contrast, do not live a sedentary life; in fact they are quite active. The most effective means of catching them is by jig when they arrive in large numbers in the protected bays of the South Coast to breed. There the eggs are laid in aggregations on the sea bed. On hatching, the young squid are carried by currents to deep water over the Agulhas Bank or even to the West Coast where they spend most of their life. Then, when they are ready to breed, the squid swim back to the bays along the South Coast. It may well be that, just like other squid around the world, most local squid die after breeding.

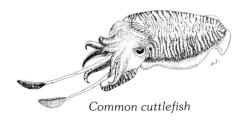

Common cuttlefish

Before the mid 1980s, almost all squid were caught commercially by trawling, but the advent of the jigging method locally has meant that many more squid are now caught annually than was the case in the past. The fishery has already had some good years and some bad ones, and whether the squid will be able to sustain such large annual catches as were made in the late 1980s remains to be seen.

The chokka squid is one of approximately 120 cephalopods known from our waters. Two of the others are illustrated on this page to show the diversity of body form in this group.

Life cycle of abalone

SPINY LOBSTERS AND PRAWNS

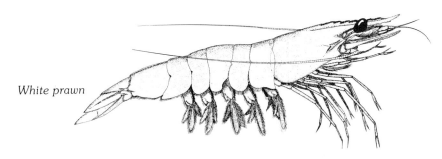

White prawn

Spiny lobster is the correct name for the group of animals you would generally call crayfish, kreef or rock lobster. It has already been shown how important spiny lobsters are to the South African fishing industry, not in terms of total mass landed, but in terms of earning value. Try comparing the price of a spiny lobster with a piece of hake in a restaurant and you will see why their importance can be gauged best in terms of value! Nevertheless, some 6 000 tons of spiny lobsters were taken annually from southern African waters throughout the 1980s, although 20 years ago the catch was greater.

There are several species living around southern Africa, two of which are shown on p. 23. Those are the two that yield the largest catches, the West Coast (inshore-living) spiny lobster or kreef and the South Coast (deep water) spiny lobster. The West Coast species is caught with two types of fishing gear. The primitive hoopnet is baited with fish and set in water shallower than about 25 m for periods of about 30 minutes. One or two fishermen in a small dinghy are all that are needed for this means of fishing. In the other method, vessels with inboard motors and larger crews set up to 75 baited rectangular traps once or twice a day in water as deep as 80 m. Whichever way they are caught, all West Coast spiny lobsters have to be a minimum size for landing. Those smaller than the specified minimum and those bearing eggs have to be returned to the sea. Much of the catch is alive on landing and that is why many of our spiny lobsters can be found in restaurant tanks in Europe and America. South Coast spiny lobsters are fished by even larger boats that set several lines of 80-120 traps, each of which are hauled once a day. Only the tails of these animals are retained. They are frozen at sea.

Perhaps one of the most interesting facts about the life of spiny lobsters is the manner in which they are thought to recolonize their preferred habitats after breeding. When the eggs hatch, they do so into flat, transparent larvae, bearing little resemblance to the adult (see the drawing on the right). These "phyllosoma" (meaning leaf organism) larvae drift in the ocean currents for perhaps several years, possibly making a full circuit of the Atlantic or Indian Ocean in doing so. Then, when they arrive back home, they metamorphose into the young form on the left, which resembles an adult, and settle onto suitable ground to grow and moult (shed the shell) regularly until they attain adulthood and can themselves breed. Surely, this must rank as one of the great survival stories of marine animals – few small marine animals travel so far in their lifetime – and it is also curious that they apparently do not colonize any other suitable territory they may pass by during their long journeys.

Adult West Coast spiny lobsters have few natural enemies other than man, and that is probably the reason why they can be so easily caught by hand as they forage for their favourite food – mussels – or any other small, slow-moving animal or scavenge material lying on the bottom.

Spiny lobsters also occur along the East Coast, but the inshore species there is very difficult to catch by hand. It rarely moves from its places of safety by day, because the animals have to avoid the attention of many more predators, (for example octopuses, sharks) than the West Coast species.

Off northern Natal and Moçambique, there are valuable fisheries for prawns. There are deep-water and shallow-water species. Inshore they have been caught with beach-seines, but nowadays most of the prawns are caught by trawlers.

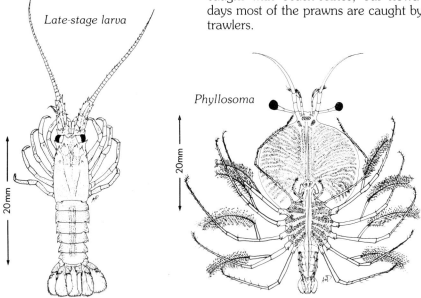

Late-stage larva

Phyllosoma

An abalone in its natural environment **1** and a shoal of chokka squid **2**, for which pigmentation plays an important part in camouflage and communication. These two molluscs grace restaurant tables. Squid are voracious predators **3** (catching a juvenile mullet), an attribute that makes them susceptible to capture with jigs **4**. The paper nautilus shell is a common find of beachcombers around the South-Western and Southern Cape, but few have seen the animal itself **5**. A curious cuttlefish approaches the photographer off the Tsitsikamma coast **6**

The two southern African spiny lobsters of greatest commercial importance are the West Coast **1** and South Coast **2** species. High densities of the West Coast species are found off Dassen Island, a generally productive fishing ground for them **3**. Their underwater habitat is sometimes strikingly colourful **4**. Even the larger decked boats are relatively small fishing vessels **5**, but at sunset at Hondeklip Bay they provide a picturesque and tranquil scene **6**

PURSE-SEINE FISHING

Purse-seine fishing is the type of fishing that yields the greatest catch off southern Africa in terms of mass. The purse-seiners, which are actually quite small vessels, sometimes come in so fully laden that the deck is almost awash with fish. If conditions are good, they may do so day after day, because the small fish that they catch are often to be found inshore close to the ports from which the boats operate.

The small fish caught by purse-seiners live near the surface in large shoals. A main purpose of grouping in shoals is to provide some form of defence against predators, but man is, of course, the greatest predator of all. He has devised a means of catching these shoaling fish better than any natural predator. When a shoal has been detected, for instance on the sonar, the purse-seiner drops a sea anchor attached to one end of the net and steams around the fish in a large circle, simultaneously paying its net out behind it. The net hangs from the surface of the water, suspended by floats. When the circle is completed, the fish are enclosed, and quickly, before the fish realize that they can escape by diving, the net is pursed. In other words, a rope passing through eyes at the bottom of the net, which has sunk because it is weighted down with lead, is hauled into the boat, and the fate of the fish is sealed. Next, much of the net itself is hauled aboard using what is known as a power block, compressing the fish into a tight mass next to the vessel. At that stage a powerful suction pump is dipped into the seething mass and the whole catch pumped aboard.

The purse-seiner then returns to port and the fish are pumped from the vessel's hold and transferred quickly into the factory for processing, perhaps into canned fish, in the case of fish such as pilchard, or into meal for agricultural purposes (for example to feed fowls), which is the fate of most anchovy.

Catches made by a single purse-seiner can be huge, upwards of 100 tons being landed at a time in good years, although the landings are nowadays generally smaller, allowing quality to be enhanced.

The main problem with the purse-seine operation is that it can be wasteful. The reasons are several. First, the skipper may not be able to identify the species he sees on his sonar and, if he is under factory instruction to catch a different species, he may waste his effort. Also, most of the fish die when they have been pursed into a tight mass and brought alongside the vessel, so even if the skipper decides to release them at that stage, the damage has been done. Sometimes too, the purse-seiner simply cannot handle the mass of fish caught, and dead fish have to be released, although attempts can be made to transfer the unwanted catch to other vessels in the vicinity. Finally, bad weather often results in wastage, as tons of fish are lost from the net or over the side of the vessel.

In times past, there was a seeming over-abundance of fish available to the purse-seiners, so to fishermen it did not seem to matter if a small quantity went to waste. Unfortunately, habits of dumping or accidental wastage are difficult to eradicate, although, fortunately, more and more fishermen are realizing that the pelagic fish resources have to be conserved carefully. Therefore, purse-seining is now being conducted more responsibly, from a resource viewpoint, and wastage is being minimized. As can be seen from the drawing, under careful management and responsible fishing practices, large catches of these shoaling fish can still be made.

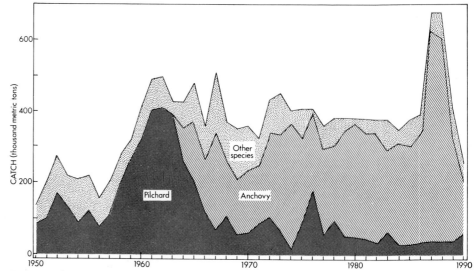

Purse-seine catches of anchovy, pilchard and other species off South Africa, 1950-1990

PURSE-SEINED SHOALING FISH

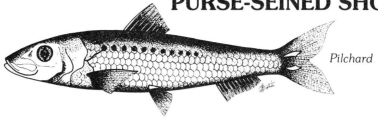

Pilchard

The two best-known shoaling fish of southern Africa are the pilchard (sardine) and the anchovy, but there are others. In the 1950s, before anchovy appeared in catches, South African purse-seine landings consisted of pilchard and horse mackerel, but the latter now only makes sporadic contributions to the purse-seine catch. In the 1960s large quantities of chub mackerel were netted, but they too are now scarce. A species at present being more commonly sought is the round herring, or red-eye as some call it. Off Namibia, pilchard, anchovy and horse mackerel have all been prominent in the purse-seine catch. Pilchard, of course, is the preferred species for canning, but round herring are also large enough to be canned. Anchovy and horse mackerel are almost totally reduced to meal and oil, although horse mackerel was canned in the years after World War II.

Both pilchard and anchovy have interesting life histories, so, to understand them better, a brief outline of what is at present known about their lives in South African waters is appropriate here. A simplified map of their life cycle is given on this page.

We start with anchovy. Adults live predominantly over the Agulhas Bank on the South Coast and spawn in large shoals throughout spring and summer. As with many marine organisms, the eggs are simply released into the water where they are fertilized and gradually carried in currents towards the West Coast. When the eggs hatch, the weakly swimming larvae are also swept up the West Coast until, if they are lucky, they arrive near the shore, group together to form shoals of young fish, feed voraciously, grow rapidly and then start actively swimming southwards, keeping close inshore. Therefore, throughout spring and summer there is a stream of eggs and larvae being swept

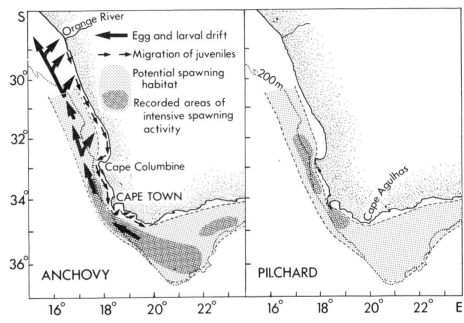

Distribution of anchovy and pilchard off South Africa

up the West Coast and then, in autumn and winter, a teeming river of young fish swimming back towards the South Coast, where the survivors will breed the next season. It is this river of juveniles that fishermen exploit as it passes the West Coast fishing ports. Some of the adults in their second or third year of life move even farther east towards Port Elizabeth, but whether their eggs join the stream being swept up the West Coast is not known. The loss of eggs, larvae and young fish is tremendous. Predators, man, failure to find food or just being swept away all contribute to the mortality. However, it only takes two eggs spawned from each female to survive to adulthood for the population to be self-regenerating.

The life cycle of the pilchard is similar in many respects. As surface currents off the West Coast generally move northwards, spawning must take place in the south for the young fish to find themselves in an area where food is most abundant when they start feeding. Then, just as anchovy do, the young fish swim southwards close to the coast. Pilchard grow larger than anchovy, live longer and spawn at an older age. Some of the older pilchard return to the productive West Coast grounds, while others migrate up the East Coast in the well-known annual "sardine run" to the coast of Natal. Just as for anchovy, the underlying force behind the success of pilchard in southern African waters is the great productivity of the waters of the West Coast. The reason for the variation in the population sizes of both species can only be speculated on, but it is probably caused mainly by fluctuations in spawning success (measured by the number of young fish surviving to join the shoals swimming southwards).

The effect of fishing by man is also a key issue, and it may combine with the dynamic environment to amplify natural fluctuations.

Views of fully-laden purse-seiners at sea **1** and approaching the harbour at Walvis Bay **2**. When catcher boats return to the processing plants, fish are removed from the hold by means of large pumps **3**. In the halcyon days of the Namibian fishery, excess catches of pilchard often spilt over the side and were wasted **4**, but the years of plenty were followed by lean years when nets would sometimes contain large quantities of jellyfish of little industrial value **5**. If there is an excess catch, other boats are nowadays called alongside to use what remains, a practice known locally as taking "bo-lyn" **6**

After the collapse of the pilchard resources, the mesh size of nets was drastically reduced to permit exploitation of anchovy. This resulted in increased catches of juvenile pilchard and horse mackerel, the other two main contributors to purse-seine catches south of the Cunene River. On a small-mesh net **1** are pilchard (top), a juvenile Cape horse mackerel (centre) and anchovy

(bottom). Purse-seine nets require frequent mending **2**. Underwater, a shoal of horse mackerel makes an impressive sight **3**. Pilchard take part in the well known "sardine run", which appears off Natal in winter, fish being cast up on the beach in large numbers **4**. When they are washed inshore, they may be gathered in the shallows **5** and are followed by predators such as sharks **6**

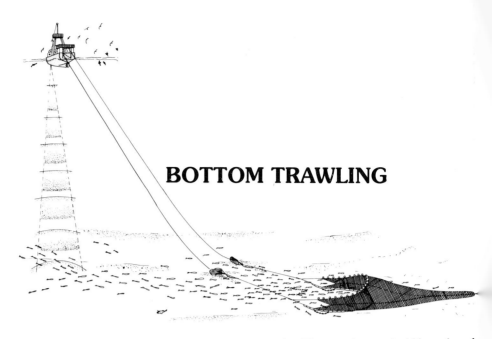

BOTTOM TRAWLING

Bottom trawling in southern Africa commenced at the end of the last century, and the manner in which it is carried out has scarcely changed to the present day. Of course, there have been improvements in technology, including the design of ships, but the way in which the trawl net is hauled along the bottom is the same now as it was 100 years ago. The species caught are the same too, although they are now caught far from the home port of the trawlers, not within sight of the houses of Cape Town, as was the tradition when trawling started there.

Some of the species sought by bottom trawlers are shown in the colour plate on p. 30, although one, the horse mackerel, is now caught mainly off Namibia by midwater trawl rather than by bottom trawl. Without a doubt, the mainstay of the bottom trawl (or demersal, to give it the correct name) fishery are the Cape hakes (which are sometimes called stockfish, Cape whiting, yankee clipper or many other trade names). Closely related to the hakes found in many other parts of the world, the Cape hakes (there are two similar species) together contribute an average of 70% of the trawl landings around southern Africa.

Hake are sought by many countries other than South Africa and Namibia, and since news of the vast potential of hake in the South-East Atlantic spread in the early 1960s, the total catch has escalated. Just take a look at the graph on this page. Until 1955, scarcely 100 000 tons of hake were being taken each year from our waters. This rocketed by the early 1970s to over a million tons. The stock clearly could not sustain this rate of exploitation and catches dropped to stabilize at a level of about 400 000-500 000 tons. Now that Namibia has declared its own 200 mile exclusive fishing zone (South Africa declared its EFZ in 1977), the catches are likely to drop even further unless that country authorizes fishing by the huge fleets which were active there in the 1980s.

Although the Cape hakes dominate the trawling industry, they are by no means the most valuable of the species caught. That attribute is held by sole, of which two species are caught, one off the West Coast and the better-tasting, but smaller, Agulhas sole on the South-East Coast. Monkfish – what a repulsive-looking fish it is (p. 30) – is also very valuable, not least because its flesh resembles that of spiny lobsters and restaurants serve it as "mock crayfish". The other traditionally sought-after fish of trawlers is kingklip, but that species has been heavily overfished recently by longlines, so its contribution to trawl catches in future will be small.

All bottom trawl species are long-lived and many of them grow to quite a magnificent size. However, being longer-lived does not guarantee their young a better chance of survival than the fry of the small surface-shoaling fish discussed earlier. As do pelagic fish, bottom fish breed by releasing large numbers of eggs into the water, and the hatchlings swim feebly near the surface for a few months before descending to the bottom to become truly demersal. Some species make daily migrations towards the surface to feed, but generally they live close to the bottom. Unlike the surface-shoalers, which often filter-feed, bottom fish tend to be opportunistic and feed on just about anything they can catch, and that often includes their own kind. In other words they are highly cannibalistic. Nevertheless, the resources of fish on the sea bed around our coast are less prone to the wild swings in abundance characteristic of the surface-shoalers. At least our mock crayfish, grilled sole and fried hake are guaranteed to be around for many years to come!

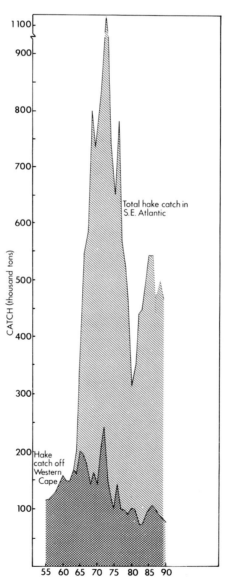

Hake catches off the Western Cape and in the total SE Atlantic, 1955-1990

LONGLINING

Longlining is the name given to the form of fishing in which lines (sometimes several kilometres long) of baited hooks are set either near the surface of the sea or close to the bottom and left for a period of some hours before retrieval. It has been a very successful form of fishing worldwide and, although it is particularly labour-intensive for fishermen, it is very selective. For example, the catch is often made up of predominantly large fish of a single species. Economically, large fish normally attract a better price than small ones per unit of mass so, provided there are fish for the longlines to catch, the returns are good. However, it is the very fact that longlines catch mainly large fish, as opposed to the mixed bag taken by nets, that has caused the demise of several longline fisheries. South Africa's bottom longline fishery is a prime example.

The bottom longline fishery off South Africa commenced in 1983 and the target fish were kingklip. Kingklip had already been taken for many years as a small by-catch in the trawl fishery for hake, but in that fishery, the kingklip catch is a mixture of small fish and large, mature adults. The total trawl catch of kingklip off South Africa rarely exceeded 5 000 tons per year (3 000-4 000 tons was more common) before the advent of longlining. Longliners immediately homed in on the aggregations of adult kingklip, especially on their spawning grounds, and the annual catch of kingklip rocketed (see the Figure). The longline catch consisted mostly of large, adult fish and many of them came from the rocky grounds over which trawlers, for fear of tearing their nets, had been unable to operate. After a massive 11 500 ton catch had been made in 1986, catch rates dropped sharply and kingklip went the way of many other overexploited fish stocks worldwide. To avoid a similar occurrence in the case of other species, particularly hake, the South African Government was forced to phase out bottom longlining and, by 1991 (a mere 8 years after it started), the fishery had all but stopped.

Surface longlining has been around for many years, but it is not South African fishermen who have been doing it; eastern nations such as Japan, Korea and the Republic of China are the main proponents of the art. The target species are almost entirely the tunas and their close relatives, all species which migrate extensively throughout the world's warmer oceans. South Africa is fortunate to lie close to the crossroads of one of these migration paths (see the Figure for longfin tuna on this page) and foreign longliners have for many years used Cape Town as a base for their South Atlantic and western Indian Ocean operations. Unfortunately, man's continuous call for more and more protein has resulted in even the tuna resources being heavily overexploited, and the discovery that gillnets can catch greater quantities than longlines has put still greater pressure on the tuna stocks. Although gillnets have been outlawed in many parts of the world (and South Africa has been particularly vociferous in condemning them), the tuna stocks are now only a fraction of what they once were. Surface longlining still does take place, but its future will obviously remain in question – another memorial to man's greed!

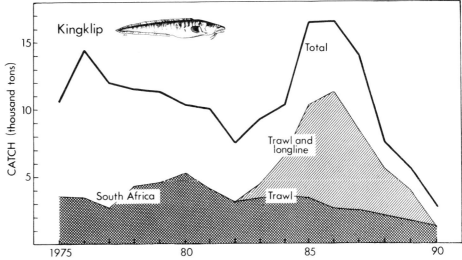
Kingklip catches in the SE Atlantic and by South Africa, 1975-1990

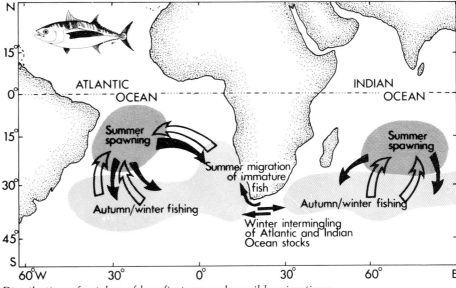
Distribution of catches of longfin tuna and possible migrations

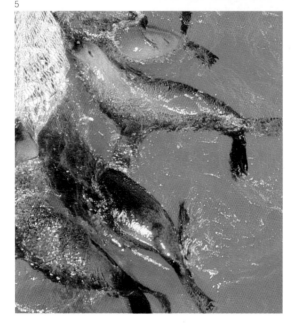

A catch on the deck of a bottom trawler off southern Africa's west coast consists mainly of Cape hakes **1**, although horse mackerel **2** and monkfish **3** are also frequently taken. Seabirds **4** (mainly kelp gulls) and fur seals **5** are species that benefit from the activities of bottom trawlers. Kingklip **6** is a highly valued catch

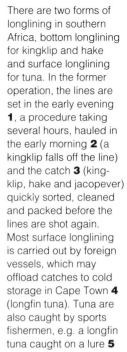

There are two forms of longlining in southern Africa, bottom longlining for kingklip and hake and surface longlining for tuna. In the former operation, the lines are set in the early evening **1**, a procedure taking several hours, hauled in the early morning **2** (a kingklip falls off the line) and the catch **3** (kingklip, hake and jacopever) quickly sorted, cleaned and packed before the lines are shot again. Most surface longlining is carried out by foreign vessels, which may offload catches to cold storage in Cape Town **4** (longfin tuna). Tuna are also caught by sports fishermen, e.g. a longfin tuna caught on a lure **5**

COASTAL FISHING

Red stumpnose

South Africa has a long and, in many places, productive, coastline. It is therefore not surprising that many people have grown up to rely on the seas close to their own doorstep not only for recreation, but also for their livelihood. Indeed there are a few settlements that subsist totally on their ability to collect protein from the sea.

You have already read that the waters off South Africa's coast fall within several temperature regimes. Natal's waters are warm, subtropical, and the home of a wide variety of fish which tend to have small population sizes. The fisherman who operates along the coast of Natal can catch or collect a wealth of different animals, although of course, by selecting his area of operation and gear carefully, he can target on just one or two preferred species. Also, the continental shelf off Natal is very narrow and much influenced by tropical water. Therefore, the coastal fisherman of Natal may see oceanic and migratory species that fall outside the range of his fellow fishermen in other parts of the country. Some residents of Natal either earn their total income, or supplement other sources of income, from commercial fishing.

In contrast to Natal, the shelf off the Southern and Eastern Cape is wider and the water temperature, although still not really cold, is temperate. The resident fauna is vastly different, although there are some species which migrate great distances or whose populations are widely distributed and can be found along much of the coastline (the Figure on this page shows the distributions of some of South Africa's popular linefish species). A number of important linefish species occur on the inshore Agulhas Bank, where numerous commercial and recreational fishermen compete for rewarding catches.

Finally, the West Coast of the subcontinent, which extends down to Cape Agulhas, is among the most productive regions in the world, primed by the upwelling described earlier. The number of fish species is less, but the population sizes of those species that do occur are sometimes huge. The West Coast is where the bulk of the commercial catch of linefish is made. It is also a Mecca of recreational fishermen. Perhaps we should say that it used to be a Mecca! Overfishing and poor management of this coastal zone has seen its potential vastly reduced, mainly during the current century, although opportunities still exist for catching galjoen from shore and snoek and tuna from boats. The infrastructure for the region's continued existence as a coastal fishing paradise is still in place, many small settlements catering largely for the fisherman (see p. 34). How many of you have seen beach-seiners hauling their nets in False Bay or have watched snoek being sold from the boats in Kalk Bay? These occasions are to be cherished and are the epitomy of coastal fishing. Management must strive to ensure that traditional fishing practices continue. Where else will we be able to obtain our mullet or bokkems?

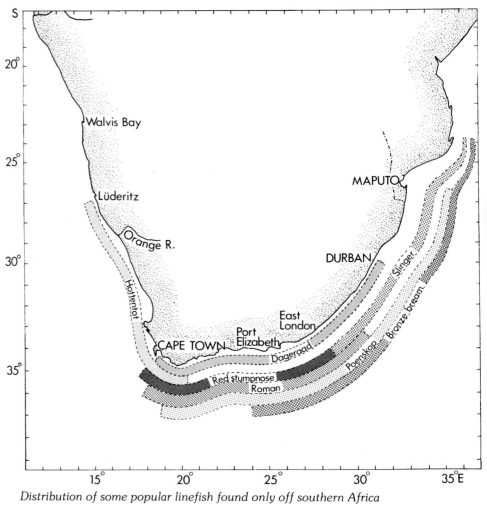

Distribution of some popular linefish found only off southern Africa

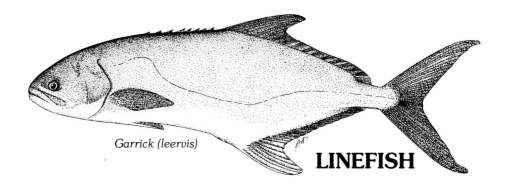
Garrick (leervis)

LINEFISH

As you learned on the previous page, linefish are caught both commercially and recreationally. Some linefish, such as snoek and tuna, are managed as specific entities because they are targeted for. However, in general a linefishery is characterized by its reliance on several species of fish.

From accounts in the literature, it is clear that linefish have been caught in South Africa since the days of Jan van Riebeeck, and probably were being caught even before that by indigenous fishermen. Historically, most of the linefishing took place around the settlements in Table Bay and Saldanha Bay, and it is likely that the catches made between the 17th and 19th centuries were excellent. Certainly, reports by people such as Simon van der Stel indicated that baited hooks dropped into the water over reefs in False Bay were quickly taken by the linefish swarming around them.

Despite the seeming abundance of fish, development of the linefishery was slow, and it was not until the British took control of the Cape Colony that several linefishing centres were established along the coastline. At first, fishing had been the domain of the local Khoi clans, but from the early 1800s, the descendants of slaves (mainly Malays) took to linefishing as a means of earning an income. Then, in the early 20th century, Cape coloureds and immigrants from Europe (notably Italians) became the fishermen and the Malays became fish hawkers. The latter situation is still prevalent today.

The 20th century has seen a notable improvement in both gear and boats, and an increase in fishing intensity. Many of the species historically targeted by linefishermen, although still

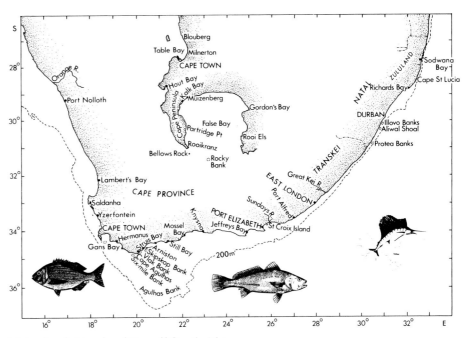
Major linefishing localities off South Africa

sought today, have become increasingly scarce. Others have grown in popularity as they have become better known.

What are the popular linefish species and where are they currently caught? It is impossible in a few lines to name more than a few of the fish species being caught. Some are shown in the colour plates and drawings on the next few pages. Many of the main fishing sites are scattered around the Western and Southern Cape, but there are also a few highly productive areas off Natal (see Figure on this page). The species being sought normally determines the fishing site preferred, for instance Dassen Island and Yzerfontein for snoek, Rooikrantz and the banks off Cape Agulhas for yellowtail, Strandfontein for kob and geelbek, and shallow inshore reefs along the coast for species such as Roman.

Species such as hottentot were being caught in False Bay in the 17th century and they are still sought by the linefishermen of Kalk Bay today. Yellowtail, now a prized linefish, were, in the early days of this century, avoided in favour of "popular" species such as geelbek. Clearly, preferences for linefish have changed over the years, but now, with catches of virtually all species becoming scarcer, the time for preferential selection is past. Conservative management alone can ensure that many species of linefish will still grace our tables in future.

A beach-seine being hauled near Simonstown **1**. Two of the small fishing harbours favoured by sport and commercial linefishermen of the southern Cape are Struis Bay **2** and Arniston **3**. Sought-after species include Roman **4**, red stumpnose **5** and hottentot **6**

A prized linefish is the yellowtail **1**, whose habit of shoaling **2** (in this instance over the Alphard Bank) has made it susceptible also to exploitation by beach- and purse-seine nets. Other species caught by line include John Brown **3**, zebra **4**, slinger **5** and musselcracker **6**

BONY FISH

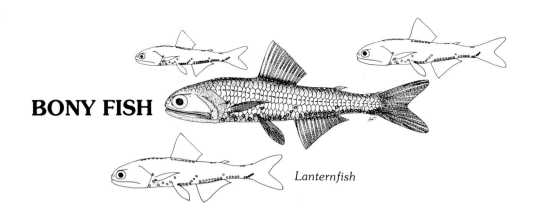

Lanternfish

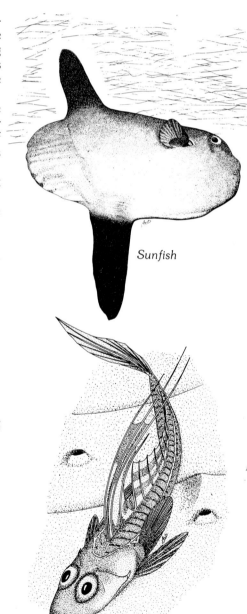

Sunfish

Monkfish *Ladder dragonet*

Bony fish are the largest and most diverse group of vertebrate animals. There are more than 20 000 species living worldwide, and southern Africa has over 2 000 of them. Many (in fact some 13%) are endemic to southern Africa, in other words they are found nowhere else. The reason for the richness of our fish fauna is probably the variety of habitats and oceanographic regimes available for them to populate. Earlier, we referred to the three very different temperature regimes around our coast, but in each of those regimes, there are habitats as widely diverse as coral reefs, estuaries, wave-swept rocky shores, sandy beaches, continental shelves and ocean depths. As stated before, the greatest variety of bony fish is to be found on the East Coast, but the West Coast is not poor in species diversity either, especially if the waters off Namibia are included. Tropical species of bony fish are regularly caught off northern Namibia.

With such a diversity of bony fish in southern Africa, it is not surprising that their shapes vary so much. Look at those illustrated on the next few pages. The lanternfish (or onderbaadjie as it is commonly known to purse-seine fishermen) may exist in quantities of millions of tons and, because it migrates vertically upwards towards the surface at night, it is sometimes caught by purse-seiners. The ladder dragonet, like the lanternfish, is a vital food organism for many predatory fish around our coast. Its colour is amazing too, its long dorsal fin being a vivid yellow. The sunfish is a predominantly oceanic fish that spends much of its time basking near the surface, whereas the attractive little seahorse is a protected species in and around Knysna.

One of our commercially sought-after bottom fish is the monkfish or anglerfish. It has a lazy existence, lying stationary on the bottom waiting for food to be attracted to its "lure". This is nothing more than an extension of its dorsal fin, which it can flick to and fro above its mouth. The monkfish blends so well into its surroundings that curious fish freely approach its lure. Watchers of aquarium specimens cannot follow the action, so quickly does the monkfish launch itself upwards to grab its prey before settling down again to enjoy the meal.

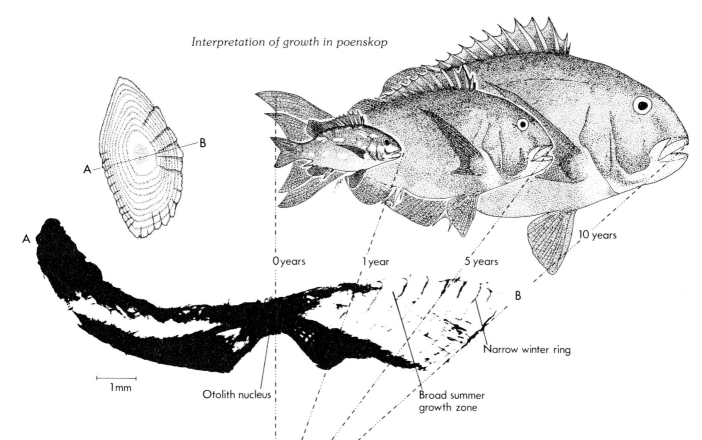

Interpretation of growth in poenskop

Obviously, each fish species is adapted for its own lifestyle, some being built for speed (for example tunas, snoek) and some, like the monkfish, for a sedentary existence. Those shown in the colour plates exemplify the colour variation in bony fish off southern Africa. Many are attractive fish. Clearly, we are lucky to be able to see these and many other species of bony fish around our coasts.

Three aspects of the life history of fish have a great bearing on their commercial and recreational importance: growth, reproduction and migration.

Growth of a fish depends on the temperature of its surroundings and the availability of food. Typically, growth is fast in young fish but gradually slows down in older ones. Generally, larger fish live longer than small fish, but this statement does not always hold true, because some large species, such as tunas and billfish, grow very fast and apparently do not live much longer than about 15 years. Indeed, most of the tuna being caught off our coasts are, despite their comparatively large size, juveniles or young adults at most. Another exception to the rule of large fish living longer than small ones concerns species of rockfish, of which we have several. One cold-water species of rockfish in the northern hemisphere attains only 46 cm but lives to over 80 years! There is always an exception to every rule!

Knysna seahorse

Age is determined by examining the rings laid down in hard structures such as bones, spines, scales or, most often, otoliths. The latter are found in the ear cavities of all fish and their prime purpose is to aid in maintaining balance. However, particularly in temperate waters where there is a seasonal temperature variation (as opposed to the tropics where the temperature tends to be stable), clear rings are laid down annually on the otoliths in much the same fashion as on a tree. They can be interpreted as shown in the Figure above, although when the fish gets older and growth (of fish and otolith) slows, the interpretation becomes more difficult, because the rings become crowded together near the margin of the otolith.

The growth rate of many of South Africa's fish species has been determined in this manner. Hake, for instance, are almost 1 m long at an age of 10 years, a kingklip 1 m long may be 30 years old, a pilchard 20 cm 2-3 years of age and an anchovy 10 cm 1 year old. Linefish tend to be slower-growing than the fish mentioned above and some growth rates of such fish are shown on p. 40.

Why, though, do we need to know the growth rate of a fish? The answer is that such information is vital to decisions made about the means of best managing a fish resource. Slow-growing fish obviously have to receive protection (for example minimum size) until they are much older than fast-growing species. Decisions on protection are based largely on the size and the age at first maturity and more is said about this on p. 40.

Among the fish found off the Southern Cape are butterfish **1**, blacktail and Roman **2**, young **3** and adult **4** red steenbras, river bream **5** and dageraad **6**

A brindle bass **1**, a tagged potato bass which remained for two-and-a-half years at the same reef off Kosi Bay **2**, a schooling coachman **3**, a red whale-fish **4**, and a shoal of strepie off Tsitsikamma **5**

Knowledge of a fish's reproductive cycle is essential in managing it. A few of our reef fish, notably the seabreams such as Roman and some of the rockcods (shown on p. 42), start life as females, but then become males with age. The advantage of such a mechanism lies in the protection that the small females receive from the larger males and the fact that females mate only with the largest males. The disadvantage is that the fishermen prefer to catch large fish, so it is the males that suffer most! However, whatever its reproductive habits, the age and size at which a fish first matures is critical, and some of the known growth rates of South African linefish are depicted in the figure on this page. Management of each should ensure that there are always sufficient fish to spawn.

Knowledge of where fish breed is important to managers and fishermen alike. The best catches will be made where the fish aggregate to reproduce. (Most fish spawn by releasing eggs and sperm into the surrounding water, so aggregation at such times is essential.) Managers, however, try to afford some protection to a species by declaring sanctuaries (for example Tsitsikamma), where fish can live and reproduce safe from being caught by man.

Many marine fish undertake some form of migration during their life. In some, for instance hake, the migration is a daily one, the fish rising in the water at night to feed. Fishermen know this and few trawlers operate at night. In other species, the migration is geographic and seasonal for purposes of spawning or to follow movements of prey species. Yellowtail and elf are among those species that migrate extensively around the coast of southern Africa. The movements of yellowtail are depicted in the second figure on this page. Again, fishermen are aware of these migrations and, to ensure an optimum catch, congregate where they know the fish will appear.

How do we know about these migratory habits? First, the very knowledge of fishermen helps to give us some clues, and then experiments with tagged fish (see p. 42) can fill in the gaps.

With all the knowledge we have about bony fish, it should be easy to protect them, but it is not. We do not yet know enough about any species, and socio-economic and recreational needs often outweigh the conservationist lobby. Also, by the time we know that a species is struggling to survive, it is often almost too late, because the stocks have already been decimated. For many of our linefish, therefore, it may well be a case of "too little, too late", but that itself is no reason not to try. Fishermen who have been lucky enough to catch a large fish, such as those shown on p. 42, will surely echo our sentiments.

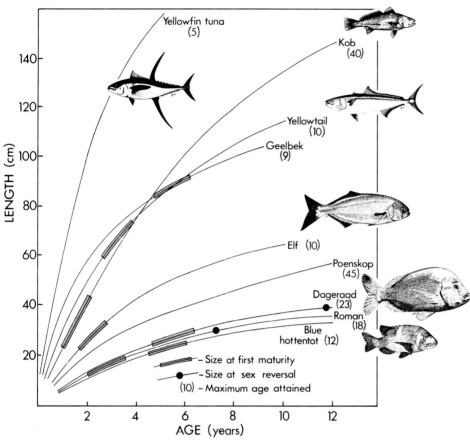

Growth rates of some South African linefish

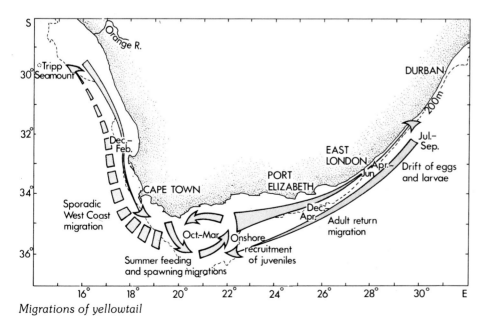

Migrations of yellowtail

SHARKS AND RAYS

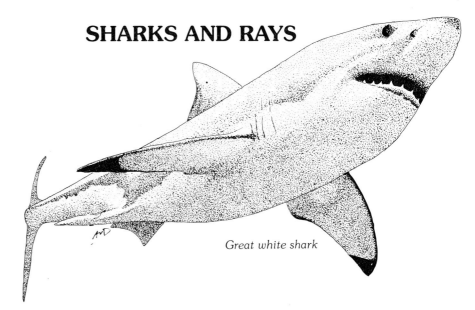
Great white shark

It is amazing the images people conjure up on hearing the word "shark". Of course, films such as "Jaws" have helped to fuel the mass fear of these animals whereas, in fact, almost all sharks are harmless. In all, 82% of shark species are less than 2 m long, and it is clearly only sharks larger than this that are potentially dangerous. Indeed, although vast numbers of people use the sea every year, fewer than 100 cases of shark attack occur worldwide, and not all those are fatal. Sharks vary in size as adults from 15 cm to well over 12 m long; the largest fish, the whale shark, is thought to attain a length of 18 m, but it is completely harmless to man. Of rays (see the bottom right plate on p. 43 – they are all flattened dorso-ventrally), perhaps only 10% reach widths of 3 m or lengths of 4 m, and none are really dangerous (stingrays can inflict painful stings, but they are rarely fatal). The next few pages will attempt to introduce a few interesting facts about these rather maligned animals.

Worldwide there are far fewer sharks and their relatives than there are bony fish. At present there are thought to be about 370 living species of shark, 500 species of ray and over 30 chimaeras. Because of its wide variety of oceanographic conditions, quite a few of them are found off southern Africa – at least 100 species of shark, some 64 rays and 8 chimaeras. Several species are endemic to our waters, but some range vast parts of the world's oceans. Of our species, two penetrate freshwater, the dangerous Zambezi shark and the largetooth sawfish. A Zambezi shark was once found as far inland as the Kruger National Park. In general, though, most of this group of fish have very narrow tolerances of temperature, salinity and depth.

Sharks, rays and chimaeras have a skeleton consisting exclusively of cartilage, the pliable material that provides support for the human nose and ears. Hence, the group is known universally as cartilaginous fish. Their skin is scaleless, in most cases covered with toothlike denticles, and they have 4-7 gill slits (rather than the bony fish's hard plate covering a gill cavity). Sharks and chimaeras use their tail to provide the thrust for swimming, whereas rays "fly" through the water by flapping their "wings", which are actually adapted pectoral fins.

The senses of cartilaginous fish are good, although sight is not always particularly well-developed. The powerful senses are those of smell, taste, hearing, vibration and electro-reception, which the cartilaginous fish use to hunt. Stories of sharks homing in on prey thrashing about in the water, of following a scent trail for long distances and of tasting prey before eating it may well be true.

The drawings on this page depict some of the manifold shapes of cartilaginous fish – all adapted perfectly to their lifestyles.

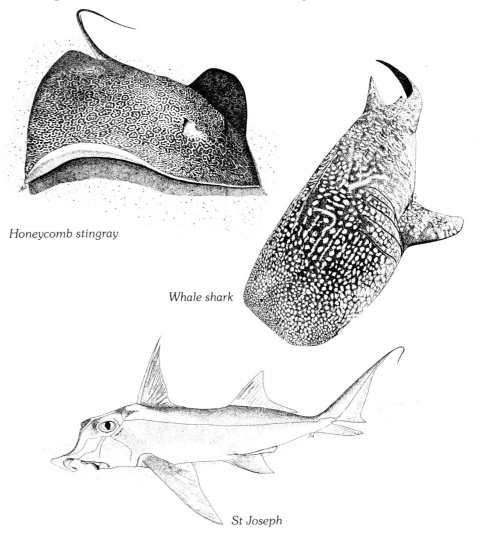
Honeycomb stingray

Whale shark

St Joseph

One species of rockcod tagged and ready for release **1** and a second seen underwater **2**. Rockcods are subject to sex reversal. In coastal reserves some fish become remarkably tame, e.g. Fransmadams **3**. A large wahoo speared off Sodwana Bay **4**, a skipjack **5** and a giant kingfish **6**

Few animals conjure up such dread in the minds of holidaymakers as sharks. However, much of this dread is unfounded because most species are quite harmless, e.g. the small Izak catshark **1**, the spotted gully shark **2** and the shortnose spiny dogfish **3**. Others are found far offshore and are seen mostly by deep-sea game-fishermen, e.g. the blue shark **4** (a specimen that has been hooked). Stingrays **5** are caught by rock-and-surf anglers but, although capable of giving a painful sting, are otherwise harmless

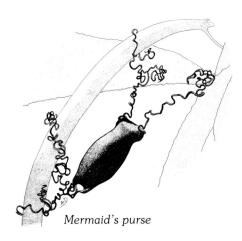
Mermaid's purse

Biscuit skate

Many beachcombers will have picked up a mermaid's purse. These are actually the egg cases of cartilaginous fish, and those you pick up usually have a small hole through which the baby shark or ray has escaped to fend for itself in the outside world. Not all cartilaginous fish produce such egg cases, however, more than half the species known to man actually giving birth to live young. Nevertheless, what is common to most cartilaginous fish is a complex reproductive procedure, often preceded by elaborate courtship. Unlike most bony fish, it is a one-on-one mating procedure, and because the young are already comparatively large when born, production (of eggs and young) tends to be very low. Unlike some bony fish, the population sizes of cartilaginous fish tend to be rather small and their potential to expand dramatically is negligible.

All cartilaginous fish are predators, but the food ranges in size from the small crustaceans preyed on by some small sharks, rays and even by the huge whale shark, to large fish, seals and other marine mammals taken by, for instance, the great white shark. Look at the teeth of the ragged-tooth shark on p. 46 and you will see that it is a fearsome predator. The sequence on the same page depicting the capture of a baby seal by the great white shark shows why that species is at the top of the food chain, fearing virtually nothing. Man is probably the single biggest danger to the great white shark, and it is therefore gratifying that, in South Africa at least, it is a protected species and cannot be targeted by fishermen. If other countries follow South Africa's lead in protecting the great white shark, its future will be close to being guaranteed.

Many South Africans are prejudiced against eating sharks and rays, but most species are highly nutritious and large quantities are consumed worldwide (though this is still only a small fraction of the quantity of bony fish that is eaten). Shark-fin soup is popular in the East and the vitamin content of shark liver is well known. For many years, a factory at Gans Bay has been producing biltong and other products from a small commercial operation, but generally speaking, the potential of cartilaginous fish off southern Africa is not really being tapped. Indeed, many spiny dogfish and biscuit skates (a ray) caught incidentally by commercial trawlers are being discarded when they actually make an excellent meal.

As fishing targets, sharks and skates are popular in competitive fishing, because points are earned for size, and most of the sharks that are caught are large. However, although there has been localized expansion of shark populations (see next page), for the reason of limited reproductive productivity it is unlikely that we will see cartilaginous fish making a greater contribution to catches. On the food table they should assume greater importance, but prejudices must first be overcome.

Tiger shark

Up to now, we have been stressing the fact that most sharks are not dangerous to man. However, there are exceptions to every rule, as experience off Natal proves. Tourism is vital to any country's and city's economy and action has to be taken to prevent any encounters which might have a long-term effect on holidaymakers' willingness to vacation by the sea. Therefore, after a recorded 21 shark incidents off Durban between 1942 and 1951, shark nets were installed there in 1952. Then, in summer 1957/'58, five people were taken by sharks along Natal's South Coast and the call for effective protection went out. Now shark nets are installed at many holiday beaches between Zinkwazi and Mzamba, as well as at Richards Bay, and the incidence of shark attack has fallen dramatically. Indeed, the numbers of sharks trapped per net have also fallen (see the Figure on this page), perhaps indicative of a declining global population or of a decreased local population, if sharks tend to be localized.

Shark attacks in South Africa have normally involved three species, the great white, the Zambezi and the tiger shark. Most attacks in Natal have taken place in murky water and many close to shore. Therefore, despite the fact that shark nets have curtailed the number of shark attacks substantially, swimming when the sea is murky should be avoided, as also should swimming at dusk and dawn, the times when feeding activity of most fish is at its maximum. It must be remembered that the shark nets do not supply absolute protection as they do not form a complete screen from surface to sea bed. In fact, 35% of the sharks caught by the shark nets are entrapped on the shore side of nets when on their way out to sea.

Protection the nets do give, but a major drawback is that they cannot be specific in what they catch. Many of the sharks and rays entrapped are not dangerous and some turtles and dolphins also fall victim to the nets. Unfortunately, despite the fact that the Natal Sharks Board checks all nets daily if possible, or at least very frequently, most of the animals enmeshed are dead by the time they are found. If they are alive they are quickly released. The Sharks Board is therefore pursuing research on the viability of other deterrents (for example electric screens). To date, however, the nets are still in place and they are likely to remain there until a satisfactory alternative is found. Some say that they have been instrumental in the increasing numbers of small dusky sharks (see the Figure), by reducing the population of large sharks that prey heavily on the small ones. The explanation may not be that simple. Other factors, such as anglers' catches, could be playing an equally influential role in such ecosystem changes. What is sure, however, is that sharks have as much right as man to exist in the marine environment. They are not all dangerous, and indeed they are a vital part of our marine ecosystems.

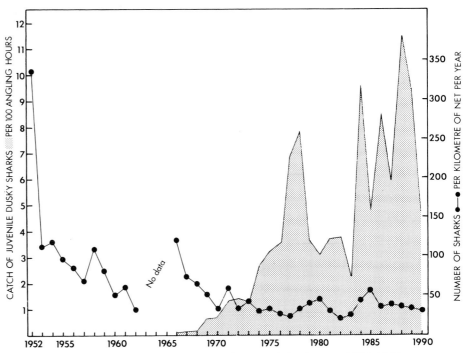

Decreasing catch rates of sharks entrapped in Natal shark nets, 1952-1990, contrasted with increasing angler catch rates of dusky sharks, 1966-1990

Shortnose spiny dogfish

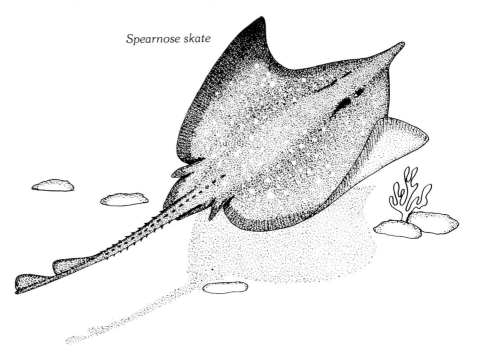

Spearnose skate

Most cinema-goers associate the word "shark" with the species that played the starring role in the film *Jaws*, but contrary to popular belief, great white sharks are not rampant man-eaters, although they are implicated in some attacks and certainly exact a toll on other marine organisms. In the sequence shown, taken near Seal Island in False Bay, it is clear that the baby seal had little chance of escaping from its predator **1-5**. A shark that is dangerous to man is the spotted ragged-tooth **6**, its fearsome appearance clear from the close-up of its head **7**.

Some more species of shark photographed underwater, a dusky **1**, another related species **2**, a blunthead **3** and a scalloped hammerhead **4**

SUBSISTENCE AND RECREATIONAL FISHING

Red steenbras

To close that portion of the book dealing with fish, we discuss some other aspects of recreational fishing and write a few words about the subsistence component of fishing in this country.

To start with recreational angling, it can be broadly divided nowadays into four distinct components, which partially overlap in the species that are caught. These are rock-and-surf angling, light-tackle boat angling, deep-sea gamefishing and spearfishing. All have growing numbers of participants (see the graph on this page which depicts the situation in Natal alone), evidence of the popularity of recreational fishing as a means of leisure. Estimates of the total numbers of South Africans engaged in the various forms of marine recreational fishing are difficult to make, but by the late 1980s there were probably over 300 000 rock-and-surf anglers, some 50 000 light-tackle boat anglers, about 13 000 deep-sea gamefishermen and perhaps 4 000 spearfishermen. Even if some fishermen are involved in more than one of these forms of fishing, the overall total still amounts to a huge number of people, and the pressure on the fish stocks to satisfy even a portion of this number of people must be tremendous.

Of course, some species have not been able to support such rates of exploitation, and the species targeted for and caught early this century are vastly different from those taken today. The drop in catch rates of some species can be ascribed to the sharing of a limited resource among an increasing number of anglers. The stocks of some species have been depleted to such an extent that they may never again contribute a significant portion of the recreational catch. There is, however, light at the end of the tunnel! Minimum size limits of fish caught have been applied for many years. Since the early 1970s many other restrictions have been imposed to protect fish and there are signs that a few species are responding favourably. Examples of these measures are: closed fishing seasons for certain species, sanctuaries (for example Tsitsikamma Coastal National Park), limitations on the number of fish caught per person each day and prohibition on the sale of some prime recreational fish species, such as galjoen.

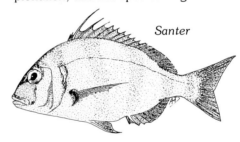

Santer

Careful management of our fish resources is essential not just for recreational fishermen, but also for the artisanal fishermen (see p. 50) and those who subsist on what they can catch in the sea. Whole communities sometimes depend on the fickle harvest from the sea, and those responsible for management have to ensure, for socio-economic reasons, a secure future for such people. However, successful management of any resource, particularly a marine resource which is by nature variable, cannot be static. There must be a dynamic process of adjustment and modification, so that the end-user accrues the fullest benefit. Fortunately, such a system is in place in South Africa today. Provided advisers continue to be listened to, subsistence and recreational fishermen may approach the future with at least some optimism.

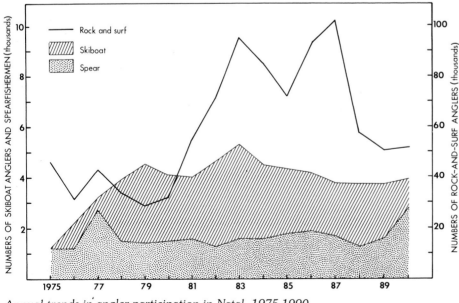

Annual trends in angler participation in Natal, 1975-1990

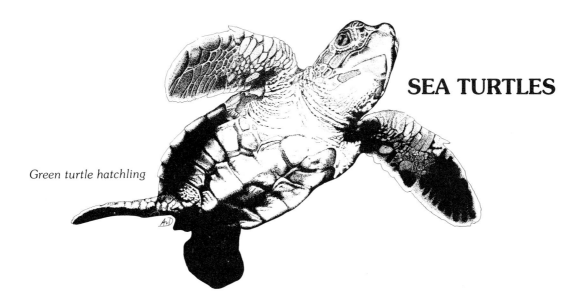

Green turtle hatchling

SEA TURTLES

Nowadays, there are five species of sea turtle which can be encountered around South Africa. Three (leatherback, loggerhead and green) are shown in colour on p. 51, and the other two are the hawksbill and the olive Ridley. Only the leatherback and the loggerhead nest on South African beaches and, although many turtles have been caught for food and their shell in the past, it is now mainly the green turtle which is farmed or harvested at some Indian Ocean islands.

Sea turtles are fascinating animals and it is likely that they have been around for millions of years; certainly they were on Earth long before the advent of man. Their life history is particularly interesting. Sea turtles have to lay their eggs ashore and it is clear that certain beaches receive preference and are returned to year after year. In fact, turtles show remarkable fidelity to their birth place. The males arrive inshore first and intercept and court the females as they in turn arrive; then they mate with them. Soon after, the impregnated females beach, usually in the early evening, and swiftly climb to a point above the high water mark. There, each female will excavate a shallow pit with her foreflippers, and ultimately a deep, flask-shaped hole in the sand with her hindflippers, which she will fill almost to capacity with eggs. She then covers the eggs with sand, compacts the nest, throws sand around to disguise it and departs to the sea. She may return as many as seven times in a season to lay eggs.

Eggs hatch after about 60 days, and the temperature of the surroundings determines whether the clutch will be totally female, all male or of both sexes. After hatching, the whole clutch scrabbles towards the surface. Generally, hatchlings only emerge at night, because once on the surface they are at the mercy of a vast number of predators, including birds, crabs and small mammals (see p. 54). They then move swiftly to the sea. Their orientation towards the sea is obviously stimulated by light, because lights on some American coastal highways attract hatchlings which are killed by passing cars. Even at night, predators are active, and 10% of the hatchlings regularly succumb before they reach the water. Even then they are not safe, but the fact that turtles have existed for so many millennia is proof that sufficient numbers do survive the swim through the predator-filled shelf zone to the comparative safety of the open ocean beyond. There they exist at the mercy of currents, feeding on such prey as jellyfish, until at a length of about 60 cm and an age of some five years, they are less influenced by currents and, in some species at least, can return to their natal beaches to breed.

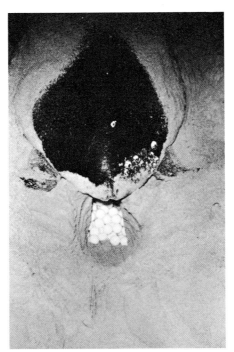 

Loggerhead turtle laying eggs (left) and turtle nesting tracks (photos G.R. Hughes)

In the Zululand estuaries, primitive traps provide protein for artisanal fishermen **1**. Elsewhere, linefishing is used for commercial **2** and subsistence **3** (a clinker boat off Jeffreys Bay) purposes. The recreational attributes of linefishing off southern Africa are well known. Some fishermen launch their skiboat through the surf at Coffee Bay **4**, and a happy fisherman displays a red stumpnose caught in False Bay **5**. Sometimes, there is scarcely a space on the beach. Fishing boats, dive boats and tow vehicles take up all available room at Sodwana Bay **6**

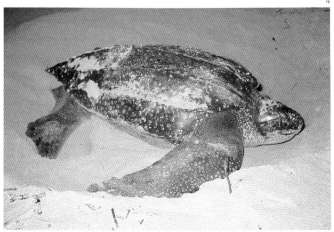

Two species of turtle breed on the beaches of Maputaland, the loggerhead **1-3** and the leatherback **4**. Green turtles **5** breed at various localities in the South-West Indian Ocean

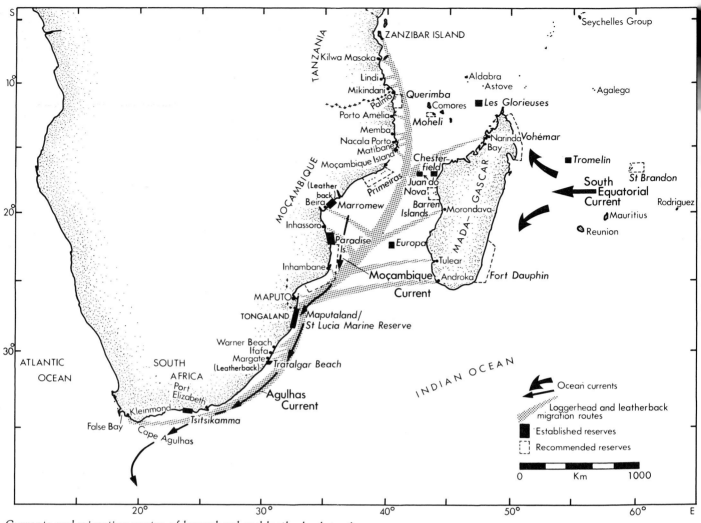

Currents and migration routes of loggerhead and leatherback turtles

Over the years, much has been learnt about the movements of sea turtles. The main source of information has been tagged animals. Tagging of turtles nesting in South Africa has been carried out since 1963, and the results are summarized in the drawing on this page.

The role of turtles in the ecosystem will never be known fully, but as a resource for man's enjoyment, they are vital. As a tourist attraction, nesting turtles are a sight not to be forgotten, and their fascinating life history is always worth reading. Beyond this, of course, they are saleable commodities, but in South Africa and on many islands of the Indian Ocean they are fully protected. Fortunately, South Africa's two

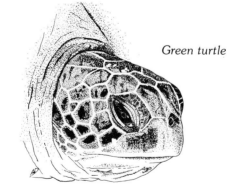

Green turtle

nesting species are not considered good eating, but the green turtle is. Fortunately again, farming of the green turtle has been shown to be economically viable. Farming, using "doomed" hatchlings which would have fallen prey to predators such as frigate birds, may yet prove a valuable conservation tool when supplemented by further controls at their nesting sites in the Indian Ocean. Many of those islands were almost totally depleted of turtles by the first few years of this century. Now, it is hoped, the future of sea turtles in this man-dominated world is somewhat rosier.

THE COASTAL ISLANDS AND GUANO PLATFORMS

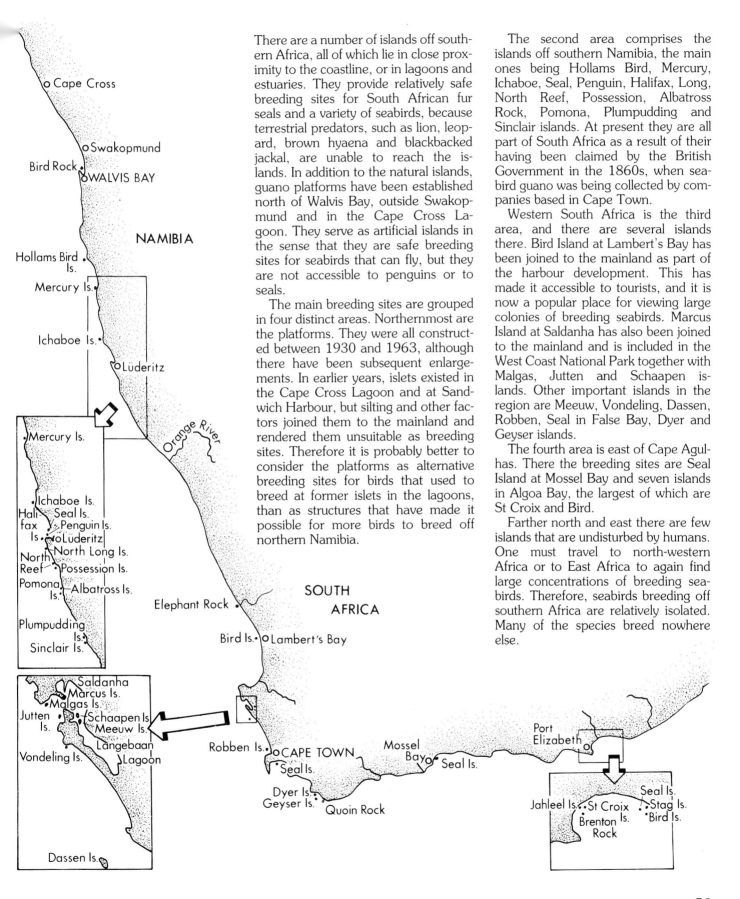

There are a number of islands off southern Africa, all of which lie in close proximity to the coastline, or in lagoons and estuaries. They provide relatively safe breeding sites for South African fur seals and a variety of seabirds, because terrestrial predators, such as lion, leopard, brown hyaena and blackbacked jackal, are unable to reach the islands. In addition to the natural islands, guano platforms have been established north of Walvis Bay, outside Swakopmund and in the Cape Cross Lagoon. They serve as artificial islands in the sense that they are safe breeding sites for seabirds that can fly, but they are not accessible to penguins or to seals.

The main breeding sites are grouped in four distinct areas. Northernmost are the platforms. They were all constructed between 1930 and 1963, although there have been subsequent enlargements. In earlier years, islets existed in the Cape Cross Lagoon and at Sandwich Harbour, but silting and other factors joined them to the mainland and rendered them unsuitable as breeding sites. Therefore it is probably better to consider the platforms as alternative breeding sites for birds that used to breed at former islets in the lagoons, than as structures that have made it possible for more birds to breed off northern Namibia.

The second area comprises the islands off southern Namibia, the main ones being Hollams Bird, Mercury, Ichaboe, Seal, Penguin, Halifax, Long, North Reef, Possession, Albatross Rock, Pomona, Plumpudding and Sinclair islands. At present they are all part of South Africa as a result of their having been claimed by the British Government in the 1860s, when seabird guano was being collected by companies based in Cape Town.

Western South Africa is the third area, and there are several islands there. Bird Island at Lambert's Bay has been joined to the mainland as part of the harbour development. This has made it accessible to tourists, and it is now a popular place for viewing large colonies of breeding seabirds. Marcus Island at Saldanha has also been joined to the mainland and is included in the West Coast National Park together with Malgas, Jutten and Schaapen islands. Other important islands in the region are Meeuw, Vondeling, Dassen, Robben, Seal in False Bay, Dyer and Geyser islands.

The fourth area is east of Cape Agulhas. There the breeding sites are Seal Island at Mossel Bay and seven islands in Algoa Bay, the largest of which are St Croix and Bird.

Farther north and east there are few islands that are undisturbed by humans. One must travel to north-western Africa or to East Africa to again find large concentrations of breeding seabirds. Therefore, seabirds breeding off southern Africa are relatively isolated. Many of the species breed nowhere else.

The most deadly predator of hatchling leatherbacks **1** in Maputaland is the ghost crab **2**. Green turtles **3** used to be farmed at Reunion, where products included spectacular turtle shells **4**. Another terrestrial predator of young hatchlings is the large-spotted genet **5**

54

Islands off the southern African coast are important breeding localities for seabirds and seals. Shown here are Possession **1**, Dassen **2**, Ichaboe **3**, Bird at Lambert's Bay **4**, Albatross **5** and Mercury **6** islands. In the Dassen Island photograph, deserted penguin nesting shelters may be seen in the foreground outside the "penguin-exclusion" wall. Both these structures were built to facilitate collection of eggs by forcing the birds to congregate at easily accessible sites

AFRICAN PENGUINS

The jackass penguin, so named because it brays somewhat like a donkey, is the only penguin that breeds in Africa. It is therefore better known as the African penguin, because some other species of penguin also have a bray-like call. Breeding only off southern Africa, the African penguin rarely ventures elsewhere, although a few stragglers have been recorded off Gabon, Zaire, Angola and Moçambique.

The African penguin is listed in the *South African Red Data Book for Birds* as "Vulnerable", a category applied to species considered likely to become endangered if factors that have caused previous population decreases continue operating. Internationally, it is regarded as a species of "Special Concern". This categorization has resulted from a massive decrease in numbers known to have taken place this century.

Penguins are not easy to census, but some indication of the extent of the decrease in numbers of African penguins can be gauged by comparing early statements with more accurate recent counts. In 1906 a lighthouse keeper estimated there to be nine million penguins at Dassen Island. In 1930 there was a report for the same island of "... no less than five million birds ...". Even if these accounts were exaggerated, there were clearly many African penguins at the island early this century. In those days the eggs of African penguins were a popular delicacy, and they were harvested in large numbers. More than 12,5 million eggs were collected at Dassen Island between 1900 and 1930, the record tally being 594 000 in 1919. As most of the eggs were usually harvested over a period of 2-3 months, it is likely that not more than two clutches from each pair of penguins was collected. The normal clutch size is two, so that there were at least 150 000 pairs of penguins breeding at Dassen Island in 1919. In 1956 there were about 72 500 pairs at the island, in 1978 some 11 220 pairs, and in 1985 only 2 757 pairs.

Estimates of the total number – both breeders and non-breeders – of African penguins at all localities are more difficult to obtain. In 1930 there were probably at least 1 200 000 African penguins, but by the mid 1950s this figure had probably dropped to some 600 000 birds. In the late 1970s there were only about 230 000, and in the mid 1980s some 160 000. In the 1980s the species was breeding at 28 localities around the coast, from Hollams Bird Island off Namibia to Bird Island, Algoa Bay, off South Africa. However, about 70% of the population was at just two localities in the southeast: Dyer and St Croix islands.

There are several reasons for the massive reduction in numbers of penguins. Decreases in the early part of the 20th century were probably caused by excessive collections of eggs, which, because of their popularity, were only terminated in 1967 when the population had already been substantially reduced. At about this time there were severe decreases in the abundance of

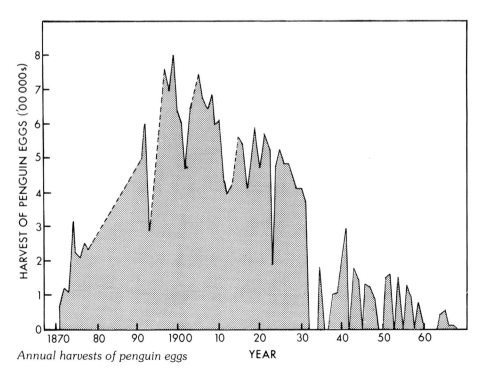

Annual harvests of penguin eggs

pilchard off both Namibia and South Africa. Being the most important food item of penguins along the West Coast, the sudden scarcity of pilchard probably resulted in poor breeding and a low survival rate of juvenile penguins. Numbers of penguins decreased dramatically at all colonies between Dassen Island and Lüderitz, but fortunately there were increases at the northern and eastern extremities of the species' range. However, these increases were not sufficient to avert a severe deterioration in the overall status of the penguin. Those colonies that have shrunk have become more susceptible to local factors. For example, feral cats at Bird Island, Lambert's Bay, and at Dassen Island are now likely to kill a far greater percentage of chicks than when many more penguins were breeding.

Mortality from oil pollution has been another factor contributing to the overall population decrease. As early as 1948, a large quantity of oil from the *Esso Wheeling*, which was wrecked off Quoin Point, washed onto Dyer Island and is estimated to have killed at least one-third of the island's penguins. Spills have continued, and many of St Croix's penguins were oiled after the *Kapodistrias* struck a reef off Cape Recife in 1985. Fortunately, oiled penguins can now be saved by treatment, and many of the affected birds were cleaned, notably at the rehabilitation station of the South African National Foundation for the Conservation of Coastal Birds (SANCCOB) near Cape Town. These two incidents highlight the susceptibility to oil spills of the two islands that support the bulk of the present-day penguin population.

Since 1970, SANCCOB has released considerable numbers of rehabilitated penguins at Robben Island. Perhaps as a result of a frequent presence of penguins in the vicinity of the island, penguins recolonized that island in 1983. Penguins had bred at the island in large numbers in the 1660s, but had ceased to do so before 1800 as a result of disturbance from humans and exploitation. In those days not only were the eggs of penguins eaten, but the unfortunate birds were themselves killed for food, for fuel to supply ship boilers, and to be rendered down for their fat. The growth of the colony at Robben Island has been spectacular. Nine breeding pairs in 1983 increased to over 1 800 in 1991. Much of the increase must have resulted from the immigration of birds hatched at other localities.

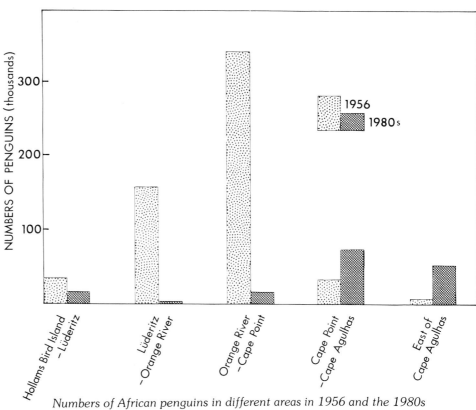

Numbers of African penguins in different areas in 1956 and the 1980s

African penguin calling

African penguins **1** are endemic to the coast of southern Africa, at Possession Island often nesting in burrows **2** or under bushes **3**. At Dassen Island, nests have been built under old guano boats **4**.

Chicks are fed on land by parents **5**, from whom they solicit food **6**. There is advantage in chicks being large **7** because when they leave for sea they will have to fend for themselves and energy reserves may tide them over the early days of inexperience

When African penguin chicks are old enough to be in no danger from predators they may be left in creches **1**, thus enabling both parents to bring food to the nest. Moulting **2** is an important part of the annual cycle. Birds must replace damaged feathers to maintain an effective insulation against

the cold seas. Moulting birds often congregate in large parties at beaches **3** (St Croix Island), where they may be counted easily. Albino penguins **4** (Dyer Island) are uncommon. Penguins that are vagrants to southern Africa include king **5**, rockhopper **6** and macaroni **7** penguins, all photographed at Marion Island

GANNETS, CORMORANTS AND PELICANS

Typical flight formation of Cape cormorants at sunset

Gannets, cormorants and pelicans are closely related to darters, frigate birds and tropic birds. Members of the last two groups occasionally visit southern African waters, but they do not breed in the region, and darters are found only in inland waters. Southern Africa has one species of breeding gannet, five of breeding cormorants and two of breeding pelicans. Together with the darter, one of the cormorants (the reed cormorant) and one of the pelicans (the pinkbacked pelican) are restricted to inland waters; a second cormorant (the whitebreasted cormorant) and the other pelican (the white pelican) occur in both marine and inland waters. The Cape gannet and the Cape, bank and crowned cormorants are restricted to the marine environment. These species breed only in southern Africa and may therefore be regarded as endemic to the region, although their migrations when not breeding sometimes take them beyond the subcontinent. For instance, juvenile Cape gannets were regularly recorded moving well up the West African coast in the 1950s. They ranged as far north as the former Spanish Sahara in those years. Along the eastern coast of Africa the Cape gannet may reach the mouth of the Limpopo River. Vagrants have been recorded as far afield as Australasia, and similarly a few Australasian gannets have been seen at South African gannetries.

White pelican

Historically the Cape gannet and the Cape cormorant were of great importance as the main producers of seabird guano, a manure much sought-after for agriculture. However, the advent of artificial fertilizers and a fall in guano harvests brought about a decrease in the value of this industry, although it nevertheless survives to the present day. A growing preference of some people in the western world to have their vegetables fertilized with natural products, coupled with imaginative marketing, could well bring about a resurgence of the guano industry. Indeed there has seldom been difficulty in disposing of guano. The coasts of south-western Africa and western South America are the only two areas where recent deposits of seabird guano are regularly harvested on a commercial basis. A few fossil deposits have been harvested elsewhere. Off western South America most of the guano is deposited by similar species – the Peruvian booby, closely related to the gannets, and the guanay cormorant. The brown pelican also deposits guano in Peru and Chile. It is more numerous than the white pelican is along the southern African coast, and is considerably more marine in its feeding habits.

For seabirds to produce commercially exploitable quantities of guano, they have to be abundant. If there is a sufficient number of birds, enough guano is deposited to make its collection worthwhile. Off south-western Africa and western South America, the seabirds contributing most of the guano feed

Guano yields off southern Africa

argely on fish, mostly small shoaling fish such as pilchard and anchovy. They tend to eat whatever fish species is abundant in the area at the time, although some prey species are preferred because they have a higher energy content than others. These high-energy species, of which pilchard and anchovy are examples, are vital components of the diet during the breeding season, when adults have to feed not only themselves but also their rapidly growing chicks.

One Figure on this page contrasts the diets of the Cape gannet, the Cape cormorant and the African penguin off northern Namibia in the late 1950s with the diets of the same three species and the bank cormorant off southern Namibia some 20 years later. One can see similarity between the diets of the four species of seabird, but also differences between regions and periods. The bank cormorant is not an abundant bird – the world population numbers about 9 000 breeding pairs. However, two islands off Namibia (Mercury and Ichaboe) are home to about 70% of all bank cormorants, and at these islands the species probably produces a fair quantity of guano, although still not as much as the Cape gannet or Cape cormorant.

The African penguin often nests under boulders or bushes where guano is difficult to collect, or in burrows in sandy soil, where the guano is soon mixed with sand. The burrows often fill with sand that is blown in, and the penguins clean their burrows to keep them open, pushing out a mixture of sand and guano with their feet. Therefore, penguin guano is of poor quality and much less valuable than that of the other birds. However, it has been collected in the past and moved to gannet breeding areas. Cape gannets construct their nests from guano and mud, but once the guano has been scraped from their breeding areas there is often not much material left for nest building. Moving in sand from penguin breeding areas therefore provides gannets with guano-rich material for nest construction.

Guano scraping has had disadvantages for birds. It has led to the areas where birds breed being lower than surrounding terrain, and heavy rain can then cause periodic flooding, killing both eggs and chicks. In 1985, Malgas Island became part of the West Coast National Park and guano has not been

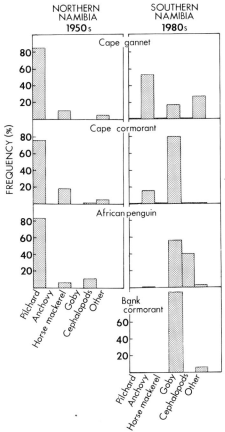

Diets of four breeding seabirds off Namibia

scraped at the island since then. Some gannets at the island now have much bigger nests than gannets breeding at other islands. The birds at Malgas Island do not need to construct new nests each year, and they start breeding about a month earlier than they used to. Therefore, the breeding season is longer than when guano was scraped. Penguins may well have burrowed into ancient deposits of guano, but at many islands the deposits were removed in the mid 1840s. The surfaces of some islands are now rocky, with few crevices for penguins to breed safe from the threat posed by the larger seals that also inhabit the islands and occasionally squash chicks and adult birds.

The guano industry has undoubtedly also had some benefits for seabirds. The construction of guano platforms north of Walvis Bay provided alternative safe breeding places at a time when the islands in nearby estuaries had become or were becoming joined to the mainland, thereby allowing access to them by predators. At some islands the area available for birds to breed has been increased by construction of wooden platforms overhanging the sea, or by the building of large walls to keep out heavy seas. Also, in former years many islands were staffed by headmen, who had to ensure that breeding birds were not disturbed by visitors to the islands. Guano was a precious commodity, and birds that produced it had to be nurtured.

Guano yields at southern African islands decreased substantially in the 1960s and 1970s, perhaps partly because artificial fertilizers were becoming widely available, but mainly because the collapse of pilchard resources off South Africa and Namibia led to less food being available for guano-producing birds. Gannets plunge from the air

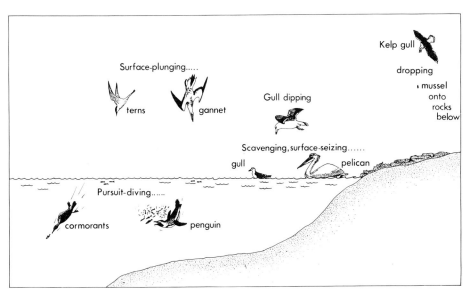

Feeding techniques of southern Africa's breeding seabirds

The Cape gannet **1** is another endemic species. It feeds by plunging onto fish at the surface **2**. The ability to fly **3** enables it to travel considerable distances, but chicks are fed at land **4**. Nests are constructed from guano, as the one occupied by a chick **5**. In the 1980s the largest colony was at Bird Island in Algoa Bay **6**. The northernmost colony is at Mercury Island **7**

Southern Africa has three endemic marine cormorants, the Cape **1-3**, bank **4-6** and crowned **7-8** cormorants. In the 1970s and 1980s, the Cape cormorant was the most numerous seabird breeding in southern Africa and the main producer of guano at platforms north of Walvis Bay

Bank cormorants

Cape gannet sky-pointing

onto fish prey near the surface of the water and search for large shoals of fish, such as those formed by pilchard. However, off central Namibia, for example, pilchard have been replaced in the food chain by the smaller bearded gobies, fish which do not congregate in such dense shoals as pilchard and which also live deeper in the water. Although other seabirds have been able to adapt to feeding on the gobies, gannets have not done so to any great extent (see Figure on p. 61).

Numbers of Cape gannets decreased from about 440 000 birds in the mid 1950s to about 260 000 in the early 1980s. Off Namibia, the food shortage has cut the gannet population by about 260 000, but this decrease has been partially offset by an increase of more than 80 000 birds off South Africa. Numbers at Bird Island in Algoa Bay have more than doubled, and this island has become the most important breeding locality for the species. In former times, Ichaboe Island north of Lüderitz supported a vast number of gannets. In 1843, the island was surrounded by more than 450 ships with crews eager to remove the guano that had accumulated down the centuries.

Cape cormorants may well have increased numerically since the mid 1950s, but the species is one of the hardest to count. It has probably replaced the African penguin as the most numerous seabird breeding in Africa. By the late 1980s, there may have been more than 850 000 Cape cormorants, compared with a population of perhaps 550 000 in the 1950s. The other cormorants are far less numerous. There are only about 30 000 adult and immature bank cormorants, approximately 8 000 crowned cormorants, and more or less the same number of marine whitebreasted cormorants around southern Africa. The only offshore breeding colonies of white pelicans are at Bird Rock platform near Walvis Bay and at Dassen Island. These two localities support about 1 000 adults, but perhaps 4 000 others breed inland at Lake St Lucia.

GULLS AND TERNS

Swift tern

Two other families of seabirds breed in southern Africa – the gulls and the terns. As do some cormorants and pelicans, a few gulls and terns also breed inland. Those breeding around the coast are kelp gull, Hartlaub's gull, greyheaded gull, roseate tern, Damara tern, swift tern and Caspian tern. Of these, the Hartlaub's gull and Damara tern breed only in southern Africa. The kelp gull and swift tern have wider distributions, but the particular forms found in southern Africa are found nowhere else. Therefore, of southern Africa's 14 breeding seabirds, nine may be regarded as endemic. This high proportion results from the isolation of suitable coastal breeding sites off southern Africa, referred to earlier.

Of the five seabirds that occur elsewhere, four can breed inland. The fifth, the roseate tern, is one of southern Africa's rarest birds. There are only about 280 adults, and these breed at islands in Algoa Bay. The species has been known to breed at Dyer Island, but not since 1971. It is one of five bird species regarded in South Africa as endangered. In Algoa Bay roseate terns breed from June to October, but then they mostly disperse elsewhere. The Caspian tern is another bird which is rare in southern Africa. There are about 200 breeding pairs, of which 100 are at Lake St Lucia. Numbers were much higher there in the past.

Most of southern Africa's seabirds breed in colonies. The Damara tern is the least colonial of all, and the only one of the 14 that breeds entirely on the mainland. It is a small tern with a total population of 1 000 – 2 000 pairs,

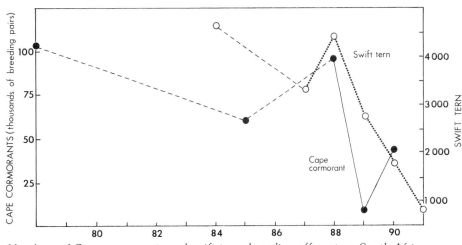

Numbers of Cape cormorants and swift terns breeding off western South Africa

the majority of which breed in summer in Namibia, north of Swakopmund. Blackbacked jackal and other predators prowl that region, and it is to the birds' advantage for their colonies to be inconspicuous. Therefore, the nests are well dispersed. When not breeding (in winter), Damara terns migrate north to the Gulf of Guinea in West Africa.

The southern African population of swift terns numbers about 5 000 breeding pairs. Many nest in mixed colonies with Hartlaub's gulls, of which there are about 13 000 pairs. Off western South Africa, the swift tern feeds mainly on anchovy, which it picks from surface waters. As shown on this page, the number attempting to breed in different years may vary considerably, probably because of variations in the quantity of food available. The greatly reduced breeding off western South Africa in the autumns of 1990 and 1991 was at a time when anchovy was scarce.

There are about 11 000 pairs of kelp gull in southern Africa, and the coastal population of greyheaded gulls is about 500 pairs – many of which breed at Lake St Lucia.

All three species of gull are opportunists and can often be seen scavenging from man's wasteful activities, for instance at rubbish dumps or behind fishing trawlers. All three have also found new safe places to breed or to roost – on the tops of tall buildings, in sewage-works and salt-works, and on guano platforms. It is the result of such adaptability that some species of gull elsewhere have been able to increase in numbers. Maybe the southern African species will follow suit, but they cannot yet be regarded as superabundant.

Seabirds of southern Africa employ a variety of breeding strategies. Some lay their eggs in spring, others in autumn, and still others virtually throughout most of the year. Kelp gulls, Hartlaub's

Whitebreasted cormorants **1-2** nest at many places along the coast as well as in inland waters. However, white pelicans **3-4** breed at only a few localities in southern Africa, one being Dassen Island. Kelp gulls **5** are one of three gulls that breed in southern Africa. Nests are scattered on the ground **6**, with the clutch normally consisting of two or three eggs. In a nest is a chick, an egg that has been cracked by a chick about to emerge, and an egg as yet intact **7**. Hartlaub's gulls are considerably smaller than kelp gulls **8**, the two species seen with crowned cormorants

The Hartlaub's gull **1** is endemic to southern Africa. Greyheaded gulls **2** also breed on the sub-continent, as do four terns: the roseate **3**, Damara **4**, swift **5** (a young bird photographed at Dassen Island, where swift terns bred for the first time in 1987) and Caspian **6** (at Dyer Island with Cape cormorants in the background) terns. In 1988, Dassen Island was the most important breeding locality in southern Africa for swift terns **7**. At some localities, migrant terns may form large roosts with species that breed locally **8**

67

gulls and African penguins are respectively examples of these three groups. Eggs produced in spring will normally have hatched by summer, when day length is longest. Because most seabirds hunt in daylight, there is more time in summer for parents to obtain food for their chicks. The Hartlaub's gull is more nocturnal than the other species. For example, after dark it has been known to catch insects attracted to street lights near Cape Town's docks. The African penguin uses the hours of darkness to swim to or from its fishing grounds.

Many of the fish preyed on by seabirds are only seasonally available in certain areas, and this fact too is an important consideration in selecting a time to breed. Bank cormorants often nest in the most precarious places, and their entire nest can be washed away by heavy seas. For species subject to such hazards, it is obviously important to be able to lay another clutch of eggs should the necessity arise, and the breeding season is then often prolonged. Of course, with an extended breeding season it is possible for parents to raise more than one brood of chicks in a year.

Greyheaded gull

Clutch size also varies. Cape gannets and swift terns generally lay a single egg, African penguins normally two, and cormorants and kelp gulls often three eggs or even as many as five. Obviously, the larger the clutch, the greater the number of mouths to feed, but the faster the potential of the population to increase. With a single chick, it is easier for the parents to ensure that it is adequately fed. Rarely do gannet chicks starve, but Cape cormorants have frequently deserted their nests if food is scarce nearby. When the nests have eggs or small chicks, one parent has to remain at the nest to incubate the egg, to shelter the young chicks from intense heat or cold, or to guard both from attack by kelp gulls, which frequently make a meal of an unprotected egg or chick. Therefore, if one parent is absent too long, the attendant parent is forced to leave to feed itself. Mass desertions have been recorded, for instance of Cape cormorants in summer 1990/91, and in that event, kelp gulls may well have enjoyed a feast.

We already know that some species are more colonial than others. Cape gannets breed in massive colonies, whereas nests of Damara terns are often so well separated that they are difficult to locate.

A number of reasons for colonial breeding have been postulated. One is that it has been forced upon seabirds by the lack of suitable nesting space at islands. Certainly this is true at islands such as Mercury, which in recent years has been recolonized by South African fur seals. The seals displaced many birds from areas where they formerly bred, including about 15% of the world's population of bank cormorants, many African penguins and many Cape cormorants. However, at some large islands there is plenty of breeding space and seabirds still breed in colonies – Possession and Dassen islands are examples.

Another possible advantage of colonial breeding is that the risk of eggs or chicks being taken by predators is reduced. At most islands there are no mammal predators, but the kelp gull and another avian predator, the sacred ibis, may well be there. Birds nesting close to each other tend to provide group protection against such natural predators. Nests near the centre of the colony may be especially safe, and experienced birds often take these sites. Less numerous species can gain protection by forming mixed colonies. Swift terns and Hartlaub's gulls often breed together. Crowned cormorants may nest in association with both these species, with African penguins, Cape gannets, egrets or herons, and even on outcrops of rock within herds of fur seals.

A third possible advantage of colonial breeding is the gaining of information about the location of good feeding areas. A nesting swift tern may, for example, observe that birds returning from a particular direction are all carrying fish whereas those flying in from another direction are doing so empty-billed.

Because they cannot fly, African penguins often do not return to colonies from the direction in which they have been feeding. They have to

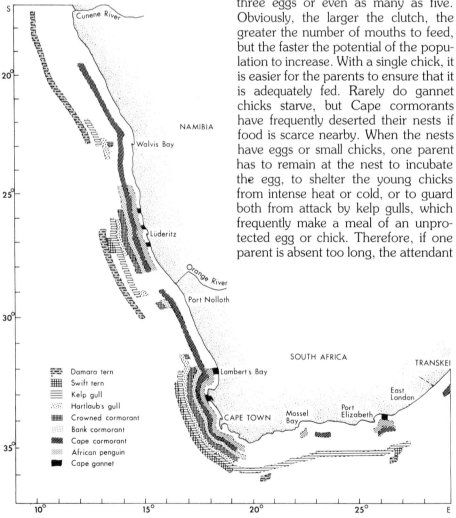
Breeding ranges of the nine seabirds endemic to southern Africa

A kelp gull at its nest

come ashore at specific landing places and then walk along paths to the colonies. For them, and perhaps also other species, the advantage of colonies could lie in the ease with which they can find a mate. Unattached birds may seek out colonies to obtain a partner and, particularly when conditions are favourable for breeding, there is clear advantage in locating a mate as rapidly as possible.

Another reason suggested as advantageous for colonial breeders is that birds huddled together can keep each other warm. This is unlikely to apply to southern Africa, though for one species of penguin breeding in Antarctica, it could be an advantage. Perhaps, however, it is just that colonies form simply because they are good places to breed – unlikely, for example, to be flooded.

A swift tern returns to a colony at Jutten Island with an anchovy in its bill (photo B. M. Dyer)

Wandering **1-2**, black-browed **3-4**, shy **5** and yellownosed **6-7** albatrosses are among the seabirds that regularly visit southern African waters

Albatrosses that sometimes visit southern Africa include greyheaded **1**, darkmantled sooty **2** and lightmantled sooty **3** albatrosses. Large groups of giant petrels often congregate around seal carcasses after sealing operations **4**. The southern giant petrel has a pale-tipped bill **5**, whereas that of the northern species is a darker red or brown **6**

MIGRANT SEABIRDS

Pintado petrel

In addition to the many seabirds that breed in southern Africa, large numbers which breed elsewhere visit the region each year. They come from inland, from the tropics, and from the cold portions of both the northern and southern hemispheres. To date, 78 species of migrant seabirds from 12 families have been recorded in southern African waters, and two others may occur at sea. This gives southern Africa one of the world's richest assemblages of migrant seabirds, although almost half the species are rare or irregular visitors to the region and therefore are not often seen.

Three penguin species occasionally visit southern Africa (see p. 59), travelling from Subantarctic or Antarctic islands. Their scarcity can be attributed to their having to swim the entire distance, whereas other birds can fly. The second family is represented by a single species, the blacknecked grebe, which is common in sheltered bays off central and southern Namibia. It breeds in southern Africa's wetlands, and so travels far shorter a distance than the other migrants. Nine species of albatross (see pp. 70 and 71), relatively large birds, have been recorded from southern African waters, the shy, blackbrowed and yellownosed being fairly common. A fourth family includes the petrels, Antarctic fulmar, prions and shearwaters, there being southern African records of 25 species (see pp. 71, 74 and 75). The pintado petrel with its spectacularly coloured plumage, the ubiquitous whitechinned petrel and the abundant sooty shearwater are particularly common. Five species of storm petrel visit southern Africa, two of which are illustrated on p. 75.

The other seven families visiting southern Africa's seas are the tropic birds (three species), boobies and gannets (three), frigate birds (one), phalaropes (one), skuas (five), gulls (six) and terns (15) – see p. 75. Additionally, there have been a number of recent sightings of the American sheathbill, but these birds are thought to have been assisted by ships in moving from the Antarctic Peninsula or adjacent islands to southern Africa. The Australasian gannet has been seen at three South African islands at which Cape gannets breed. Many of the skuas, gulls and terns are found inshore, some of the terns (for example sandwich and common) roosting in mixed flocks with species that breed locally (see p. 67).

Most of the birds that breed in the far north or far south of the globe do so in summer, when long hours of daylight increase primary production and hence food availability, providing parents with more time to obtain food for their young. They migrate away from their breeding grounds mainly in winter. Therefore, birds from the cold portion of the northern hemisphere tend to be seen off southern Africa in the northern winter, in other words the southern African summer. Conversely, migrants from the Southern Ocean visit southern Africa mainly during the southern winter. The greatwinged petrel is an exception. It breeds at Subantarctic islands in winter and is most abundant off southern Africa during summer. Of course, only breeding birds are tied to breeding localities by the responsibilities associated with rearing chicks. Many nonbreeders remain at sea off southern Africa throughout the year.

Great shearwaters (photo P.E. Malan)

Common tern (photo P.E. Malan)

Migrant seabirds visit southern Africa for a number of reasons. Some of the rarer ones probably arrive unintentionally, perhaps being swept off course by strong winds or currents. Other species merely pass through on their way between breeding and non-breeding areas, examples being the blackbellied storm petrel and the Arctic tern. The former breeds in the Southern Ocean and winters in the tropical Atlantic, Indian and Pacific oceans. Arctic terns breed in the northern hemisphere, but winter in the Southern Ocean along the Antarctic continental shelf. Such passage-migrants are most abundant off southern Africa during their outbound and return migrations, in other words in autumn and spring.

Wandering albatross

Around southern Africa there is a plentiful supply of food, and many of the migrant birds visit the region during their non-breeding season to feed. They employ a variety of techniques to catch prey. Some locate it from the air and plunge onto it – boobies, terns and tropic birds are examples. Others, such as albatrosses, phalaropes, petrels and gulls, may seize prey while sitting on the surface. Smaller birds, for example storm petrels, prions and terns, often snatch prey from the surface while running along it or flying. Penguins, grebes and some shearwaters chase prey underwater, attaining rapid speeds. The various species are not necessarily restricted to one method of feeding. For example, albatrosses, petrels and some storm petrels can also dive beneath the surface.

Many birds benefit from man's fishing activities, which are by nature wasteful. Fish spill from purse-seine nets or escape from bottom or midwater trawls. From the trawl fisheries too, unwanted fish may be discarded together with unwanted portions of sought-after fish, such as hake heads and viscera. Such discarding provides a ready supply of food at the surface of the ocean for birds able to take advantage of it. Blackbrowed and shy albatrosses are abundant at trawlers, as are whitechinned and pintado petrels, but other albatrosses and petrels, skuas, shearwaters, storm petrels, prions, terns and gulls may also feed at trawlers. Two seabirds which breed locally, Cape gannets and kelp gulls, often join the flocks of birds feeding behind trawlers. In sealing operations, it is often only the pelt of the animal that is sought and the rest of the carcase is sometimes thrown into the sea. Northern and southern giant petrels use their powerful bills to feed on seal carcases, as shown on p. 71, and they also used to gather at whaling stations at Langebaan and Durban Bluff when South Africa caught whales.

Skuas and frigate birds obtain their food by chasing other birds and forcing them to regurgitate, a practice termed klepto-parasitism (*kleptes* = thief). Terns are often victims of such a practice, although larger birds such as gannets may also be harassed.

Australasian gannet in front at right (photo B.M. Dyer)

Blackbrowed albatross and Subantarctic skuas at a trawl net (photo J. Gates)

Pomarine skua harrassing Cape gannet (photo P.E. Malan)

The Antarctic fulmar **1**, Antarctic petrel **2**, pintado petrel **3**, soft-plumaged petrel **4**, broadbilled prion **5**, fairy prion **6**, grey petrel **7** and whitechinned petrel **8** visit southern Africa. The Antarctic petrel is a very rare vagrant, the fairy prion a rare vagrant

Other migrant seabirds include Cory's **1** and sooty **2** shearwaters, Wilson's **3** and blackbellied **4** storm petrels, the pomarine **5**, Subantarctic **6** and South Polar **7** skuas, Sabine's gull **8** and Antarctic **9** and Arctic **10** terns

SEALS

Crabeater seal on ice floe

Down the years the South African fur seal has attracted widespread attention – from sealers, fishermen and conservationists. It is the only seal that breeds in sub-Saharan Africa, and it is closely related to the Australian fur seal, which is a different form of the same species. The world has 34 species of seal, which can be distinguished in three families – the "true" seals (19 species), the "eared" seals (14 species) and the walrus (one species).

The "eared" seals include the fur seals and sea lions. They are exclusively marine, and have small external ears, which are rolled to exclude water, and hindflippers, which can be turned forward to act as legs for walking on land. They use the large foreflippers for swimming, and the hindflippers mainly for steering. They have no hair on the palms or soles of their flippers and, under their hair, their skin is pale. "True" seals live in both freshwater and the sea, they have no external ears, their palms and soles are hairy and the underlying skin is dark. Their hindflippers cannot be turned forwards, and they are consequently less mobile on land than the "eared" seals.

The "true" seals include the southern elephant seal, the leopard seal and the crabeater seal, all of which occasionally visit southern Africa, as does another fur seal – the Subantarctic fur seal. The southern elephant seal and the Subantarctic fur seal breed at Subantarctic islands and are seen in southern Africa more often than the leopard and crabeater seals, which are truly Antarctic species. However, even the Subantarctic species travel enormous distances to reach southern Africa. Most southern elephant seals probably come from South Georgia, which is about 5 000 km away.

Male southern elephant seals attain a huge size. They may measure between 4 and 5 metres from nose to tail, and weigh up to 3,6 tons, making them the world's largest seal. The females are considerably smaller. It is estimated that there are over half a million southern elephant seals, with South Georgia supporting about half the total number. The Subantarctic fur seal is an attractive animal, having a cream-coloured chest and face (see p. 79). The total population is estimated at more than 200 000 animals, about 90% of which are at Gough Island. The crabeater seal is the world's most abundant seal, numbering in the millions. It feeds almost exclusively on krill. These crustaceans live in large swarms and they can be sucked into the mouth and sieved from the water through spaces between the cusps of elaborate cheek teeth. Leopard seals are voracious predators, eating penguins and other birds as well as other species of seal, including the crabeater seal. Leopard

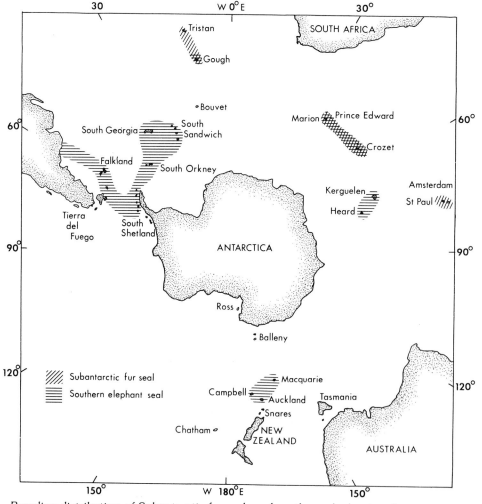

Breeding distribution of Subantarctic fur seals and southern elephant seals

seals are less numerous than crabeater seals, but are nevertheless plentiful – conservative estimates have ranged between a quarter of a million and 800 000 individuals.

The South African fur seal is another abundant seal, although there have not always been as many as there are now. It was exploited even before Europeans began settling in southern Africa in the 17th century. Harvesting was without legal control until late in the 19th century, when there were probably considerably fewer than 100 000 left. Regulations were then implemented and, although more than 2,5 million pups and bulls are known to have been harvested between 1900 and 1983, there were estimated to be more than a million South African fur seals at the start of the 1990s. This represents a remarkable recovery which can be regarded as one of the outstanding success stories of marine conservation in southern Africa.

South African fur seals give birth to pups at islands or on the mainland coastline. The present-day distribution of breeding colonies, and of non-breeding aggregations, is shown on the accompanying map. There is evidence that seals formerly bred at 19 islands where they now no longer do. In fact, exploitation by man and its accompanying disturbance led to seals abandoning 23 islands, of which four have since been recolonized. At the time of European settlement of southern Africa, seals probably did not breed on the mainland, where they would have been hunted by coastal peoples and predators. Numbers of larger predators, such as lion, were soon greatly decreased and, when it was realised that some barren coastal stretches were rich in diamonds, human access to these areas was restricted. This made some mainland coastal sites suitable for colonization by seals and there are now far more seals breeding on the mainland than at islands. Terrestrial predators have not been totally eradicated, and two that benefit from the present breeding by seals on the mainland are the ubiquitous blackbacked jackal and the rare brown hyaena.

South African fur seal pups are suckled by their mothers for a period of 8-10 months. The pup is entirely dependent on milk for the first six months of its life, but thereafter it begins to supplement its diet by hunting for food, such as mudprawns, crabs, small spiny lobsters and small fish, in inshore waters. Bulls play no part in the rearing of pups and, to maintain an adequate supply of milk, females leave their pups for a few days at a time, returning to sea to feed. It is on such occasions that pups are particularly susceptible to predation by the mainland predators. On returning from feeding trips, cows must recognize their pup from among the thousands there may be at a breeding colony. She calls loudly as she arrives, and her pup often recognizes the call and dashes towards her. On her part, she can distinguish the pup by its call and its smell.

Seals are perfectly adapted for existence in water. South African fur seals have been known to dive to depths of 200 m and to remain underwater for as long as 7½ minutes. Elephant seals have been recorded diving to 900 m, and Weddell seals frequently remain underwater for 15-30 minutes. As seals are air-breathers, they must carry large quantities of oxygen on such dives. This is ensured by seals having large volumes of blood, with high levels of haemoglobin, the normal oxygen-binding molecule, in it. Additionally, seal muscles have a high concentration of a closely related molecule, myoglobin, which also binds oxygen.

The rapid increase in numbers of South African fur seals during the 20th century, and the large numbers now breeding on the mainland, pose the interesting question of how the present-day abundance compares with what existed prior to the arrival of Europeans on the subcontinent. At some localities, for example the islands off Namibia, it is possible to map former distributions

South African fur seal pup

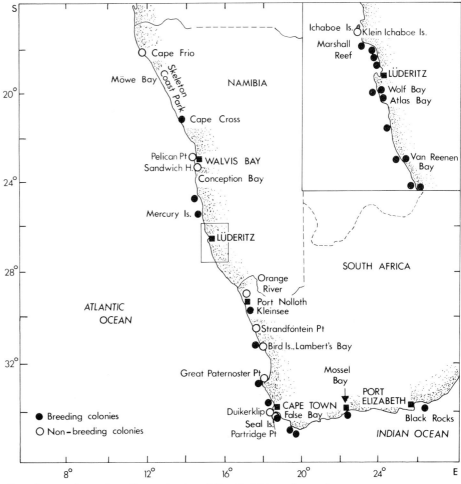

Breeding and non-breeding colonies of South African fur seals

The South African fur seal is at home in water **1**. Bulls are much larger than cows, a discrepancy especially noticeable in mating **2**, when bulls aggressively defend their territories **3**. Pups **4** are suckled by their mothers **5** until weaned at an age of 8-10 months. When they wander too far away from their mothers they are retrieved **6**

Pups of South African fur seals are sometimes preyed upon by black-backed jackals **1**. Twins **2** are rare in nature and it is uncommon for more than one to survive. Also seldom seen are albino fur seals **3**. Rare vagrants to southern African waters include the Subantarctic fur seal **4**, its cream-coloured face and chest conspicuous on a photograph taken just north of Sandwich Harbour, the southern elephant seal **5** and the leopard seal **6**

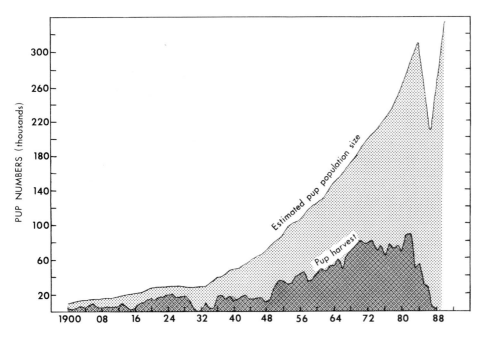

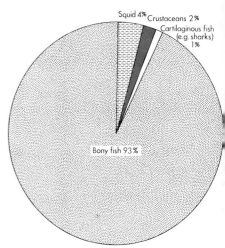

Diet of the South African fur seal

Numbers of South African fur seal pups born and harvested, 1900-1989

of seals from remains, such as fur and bones, that still exist in the soil. This information suggests that, in some regions at least, seals may be approaching their former levels of abundance.

Harvesting of seals has become a controversial issue. The most valuable product has generally been the skins of pups aged 7-10 months, which are highly suited for processing to items such as soft, luxurious ladies' coats. These were sold in the fashion stores of Europe and North America, but during the 1970s and 1980s, conservationist lobbies urged countries not to import seal products, and the traditional markets for seal skins collapsed. Many people regard the method used to kill seal pups as particularly abhorrent. Pups are first clubbed on the head to render them unconscious, and then the blood vessels around the heart are slashed with sharp knives. The humane dispatch of seals requires both skill and experience; yet, in some recent operations inexperienced and untrained personnel have been used. Is this all necessary, when it is borne in mind that many of the products are non-essential, luxury items?

On the other hand, there are the fishermen who assert that the large numbers of seals are prejudicing their catches. Some seals take fish from lines; others jump into nets to feed on fish trapped there. Certainly, a million seals require a lot of food to survive. It is estimated that they consume more than a million tons of food each year, and of this more than 60% is made up of species also sought by the commercial fishing industry. If the number of seals continues to increase, it can be expected that they will eat even more fish.

Seal sharing fish with other predators

CETACEANS

Layard's beaked whale

Cetaceans are warm-blooded mammals that breathe air and give birth to live young, which the mother suckles on milk from her mammary glands. Their bodies are streamlined for efficient movement through water, so they have few external protrusions and an absence or scarcity of hair. Insulation is provided by a layer of fat, known as blubber, which in the bowhead whale of cold Arctic waters may be as thick as 50 cm. There are two major groups of cetaceans, the toothed whales (Suborder Odontoceti – including sperm whales, beaked whales and dolphins) and the baleen whales (Suborder Mysticeti – including right whales, pygmy right whales and rorquals).

Worldwide, the order Cetacea includes some 79 species, most of which are marine, although a few species may be found in large rivers such as the Indus and Ganges of the Indian subcontinent, or the Amazon and Orinoco of South America. Slightly less than half of the world's diversity, 37 species, has been recorded off southern Africa. These species are listed in the table on this page along with information on their maximum lengths and their distributions off southern Africa. The large number of cetacean species found off southern Africa nearly equals totals for the entire North Atlantic or North Pacific oceans (40 and 41 species respectively). This can be attributed to the wide range of oceanographic conditions found off southern Africa, ranging between warm temperatures in the east and cool water in the west.

Some species, for example Arnoux's beaked whale and Hector's beaked whale, are known in southern Africa from a few strandings only – specimens that have washed out on beaches. There have been no confirmed sightings of these two animals at sea, and they can best be considered vagrants to the region. Very little is known about most of the beaked whales (family Ziphiidae). They are often thought to be creatures of deep waters and to feed at great depths, sometimes on the ocean floor, mostly on squid. Some

WHALES AND DOLPHINS OF SOUTHERN AFRICAN WATERS

Common name	Scientific name	Maximum length (m)	Distribution*
BALEEN WHALES	Suborder Mysticeti		
Right whales	Family Balaenidae		
Southern right whale	Eubalaena glacialis	15,0	C; O; SW – SE
Pygmy right whales	Family Neobalaenidae		
Pygmy right whale	Caperea marginata	6,5	C; O; W – SE
Rorquals	Family Balaenopteridae		
Blue whale	Balaenoptera musculus	30,0	O; W – E
Fin whale	B. physalus	27,0	O; W – E
Sei whale	B. borealis	21,0 (female) 17,5 (male)	O; W – E
Bryde's whale	B. edeni	15,0	C; O; W – SE
Minke whale	B. acutorostrata	10,5	O; SW – E
Humpback whale	Megaptera novaeangliae	16,0 (female) 15,0 (male)	C; O W – E
TOOTHED WHALES	Suborder Odontoceti		
Sperm whales	Family Physeteridae		
Sperm whale	Physeter macrocephalus	18,0	O; W – E
Pygmy sperm whale	Kogia breviceps	3,5	O; W – E
Dwarf sperm whale	K. simus	2,7	O; SW – SE
Beaked whales	Family Ziphiidae		
Southern bottlenose whale	Hyperoodon planifrons	7,5	O; SW – SE
Arnoux's beaked whale	Berardius arnuxii	9,5	O; vagrant?
Cuvier's beaked whale	Ziphius cavirostris	7,0	O; SW – SE
Layard's beaked whale	Mesoplodon layardii	6,0	O; W – S
Blainville's beaked whale	M. densirostris	4,5	O; SW – E
True's beaked whale	M. minus	5,0	O; SW – SE
Gray's beaked whale	M. grayi	5,5	O; S – SE
Hector's beaked whale	M. hectori	4,5	O; vagrant?
Dolphins and their kin	Family Delphinidae		
Killer whale	Orcinus orca	7,0 (female) 9,5 (male)	C; O; W – E
False killer whale	Pseudorca crassidens	5,0 (female) 6,0 (male)	O; W – E
Longfinned pilot whale	Globicephala melaena	5,5 (female) 6,0 (male)	O; W – SE
Short-finned pilot whale	G. macrorhynchus	4,0 (female) 5,5 (male)	O; S – E
Pygmy killer whale	Feresa attenuata	2,7	O; W – E
Melon-headed whale	Peponocephala electra	2,7	O; vagrant?
Risso's dolphin	Grampus griseus	3,5	O; W – E
Bottlenose dolphin	Tursiops truncatus	3,9†	C; O; W – E
Common dolphin	Delphinus delphis	2,5	C; SW – SE
Striped dolphin	Stenella coeruleoalba	2,7	O; SE – E
Spotted dolphin	S. attenuata	2,5	C; O; SE – E
Spinner dolphin	S. longirostris	2,0	C; O; E
Fraser's dolphin	Lagenodelphis hosei	2,6	O; SE – E
Southern right whale dolphin	Lissodelphis peronii	2,4	O; W
Dusky dolphin	Lagenorhynchus obscurus	2,1	O; W – SW
Heaviside's dolphin	Cephalorhynchus heavisidii	1,7	C; O; W – SW
Rough-toothed dolphin	Steno bredanensis	2,8	O; vagrant?
Indo-Pacific humpback dolphin	Sousa plumbea	2,8	C; S – E

* C = coastal, O = offshore
  Divisions between West (W), South-West (SW), South-East (SE) and East (E) coasts have been made at Table Bay, Cape Agulhas, Algoa Bay and Cape St Lucia respectively
† SE Coast form – 2,6 m

Sealing has been a matter of some controversy. In the pup harvest, groups of pups aged about nine months are herded together **1** and then in small numbers forced to run the gauntlet between lines of men armed with clubs **2**. After clubbing "sticking" occurs, when sharp knives are used to slash around the heart of the pups **3** (foreground). The blood is hosed away, and pups are then skinned **4**. Carcasses are loaded onto trucks for burial **5**. Skins have the blubber removed, a process known as beaming **6**, and are then salted to dehydrate them **7**. Bulls may be shot and dragged up the beach **8**. Sometimes they are skinned and their flesh used to produce meat meal

Heaviside's dolphins are endemic to southern Africa's west coast. They occur inshore and frequently bow-ride next to boats **1**. They occasionally breach **2-4**. Few photographs exist of this small dolphin, but two animals were photographed off the Groen River at a depth of 30 m **5**. At the surface, the characteristic triangular dorsal fin and the pale grey cape at the front of the body are readily apparent **6**

Southern right whale dolphin

beaked whales may live for 70 years or longer.

The dolphins, a family (Delphinidae) that includes the awe-inspiring killer whales, are more readily seen. One of these, Heaviside's dolphin, is found only in the Benguela upwelling system off western southern Africa. Groups of Heaviside's dolphins are often encountered at specific localities, for example in Hottentot Bay or near Lüderitz on the Namib coast, and near Port Nolloth and Paternoster off western South Africa. At such places they frequently approach boats and "bow-ride" – travel just in front of a boat's bows – providing excellent opportunities for viewing.

A second dolphin species frequently observed off western southern Africa is the dusky. It, too, is a relatively small dolphin and a southern hemisphere species. Apart from southern Africa, it also occurs off southern Australasia, southern South America, and in the vicinity of Kerguelen Island in the southern Indian Ocean. It is a sociable animal, often seen in groups, which may at times number in excess of 100 individuals. Duskies frequently leap from the water, and a few have survived for many years in oceanaria. The keeping of dolphins in captivity is a controversial issue, but it has greatly increased knowledge about them and has contributed to changed attitudes to some species – for example, killer whales were once regarded with fear and hostility, but are now often the objects of affection.

Two other species with a cool-water distribution are the southern right whale dolphin, so named because it resembles right whales in lacking a dorsal fin, and the long-finned pilot whale. Both have penetrated the cool Benguela system on the West Coast. The southern right whale dolphin is spectacularly marked in black and white. The few privileged to have seen it are unlikely to forget the experience.

The bottlenose dolphin is widespread in tropical and temperate oceans of the world. As with duskies, bottlenose dolphins have been maintained in South African aquaria, where they have delighted many with their acrobatic performances. Bottlenose dolphins are reasonably plentiful off southern Africa's west coast, but they also live along the southern and eastern seaboards. Those found on the eastern shores attain a smaller size than bottlenose dolphins on the West Coast. On the East Coast, this species lives close inshore, a factor that makes it relatively easy to see, but that also makes it susceptible to capture in the shark nets set along Natal's beaches. Unfortunately, most of the dolphins that become entangled in the nets drown.

The common dolphin has a similar world distribution to that of the bottlenose dolphin, but it is seldom encountered north of Cape Columbine on southern Africa's west coast. Elsewhere it is relatively abundant, in some areas seasonally so. It may be seen off Natal and Transkei in winter, attracted there by the abundant food provided by Natal's annual "sardine run". As with bottlenose dolphins, several hundred common dolphins have died in Natal's shark nets. Bottlenose and common dolphins also sometimes live in large schools. Such schools probably serve a number of functions, including protection from predators, helping young animals to learn, and increasing feeding opportunities – schools of dolphins may be able to herd prey until each member has fed. The toothed whales that form the largest schools feed predominantly on prey that itself is frequently found in dense aggregations, such as pilchard and lanternfish.

Three other species that are widespread off southern Africa are Risso's dolphin, killer whale and false killer whale. Of these, Risso's dolphin is

Male killer whale

thought to be the most abundant. Killer whales are large animals found over most of the world's oceans. Adult males have a particularly tall dorsal fin, which enables them to be distinguished from females and immature animals. Killer whales eat a wide variety of prey, including squid, sharks, bony fish, seabirds, seals and other cetaceans.

The warm waters of the Agulhas Current allow five species of dolphin with tropical or subtropical distributions to reach South Africa's east or south coasts – Fraser's dolphin, Indo-Pacific humpback dolphin and three dolphins of the genus *Stenella*, the spinner, spotted and striped dolphins. Large numbers of spinner and spotted dolphins have been captured and drowned by purse-seine nets fishing for tuna in the Pacific Ocean. These two dolphins feed on small fish and squid, as do tuna, which probably leads to the dolphins and tuna associating with each other. Altered fishing techniques have reduced the incidental mortality of the dolphins, but it still remains a problem. The Indo-Pacific humpback dolphin is a coastal species, rarely found more than a kilometre offshore in South Africa. It eats cuttlefish and fish typical of turbid conditions or estuaries, for example mullet, and is another dolphin to have fallen victim to Natal's shark nets.

The third family (Physeteridae) of toothed whales to be found off southern Africa is the sperm whales, which range in size from the dwarf sperm whale (maximum length 2,7 m) through the pygmy sperm whale (3,5 m) to the sperm whale (18 m).

The sperm whale itself is the largest of all the toothed whales, which distinction earmarked it as a target for commercial whalers. The pygmy and dwarf sperm whale seem to prefer relatively warm water, but the sperm whale migrates between the equator and cold polar water. The sperm whale has an enormous head which contributes about a third of its total body length. It has an underslung lower jaw with large teeth, and adults may consume up to a ton of squid, including giant deep-sea squids, each day. However, the deep-sea squids constitute a relatively small portion of the total diet. Sperm whales also eat octopus and a variety of fish. They may dive deeper than 1 000 m and occur singly or in groups of more than 50 animals. The larger herds may be bachelor groups, schools of immatures or schools of females and juve-

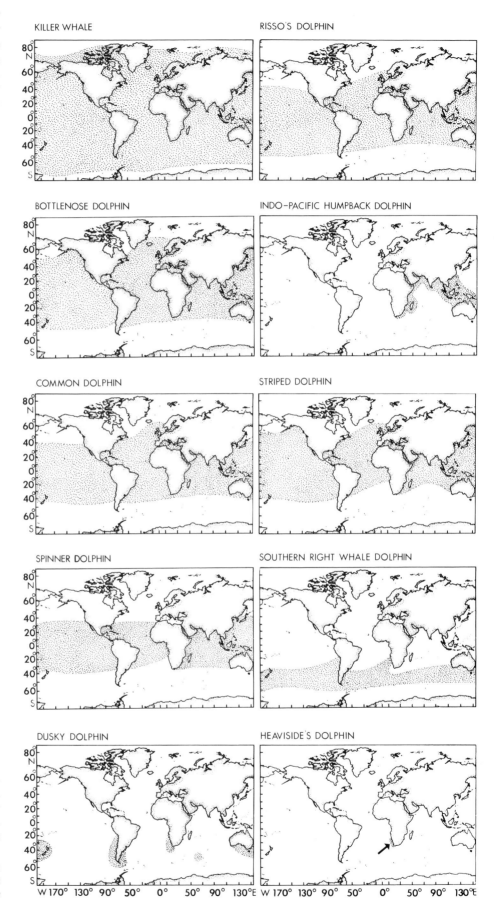

Worldwide distribution of 10 species of dolphin encountered off southern Africa

Dolphins seen off southern Africa include false killer whales **1**, killer whales **2-4**, long-finned pilot whales **5** and dusky dolphins **6-7**. Killer whales are predatory cetaceans. That shown in **4** was photographed with humpback whales and has what is probably a piece of flesh from one of these whales in its mouth

Some other southern African cetaceans are the common **1-2**, bottlenose **3-4**, spotted **5**, striped **6-7** and spinner **8** dolphins

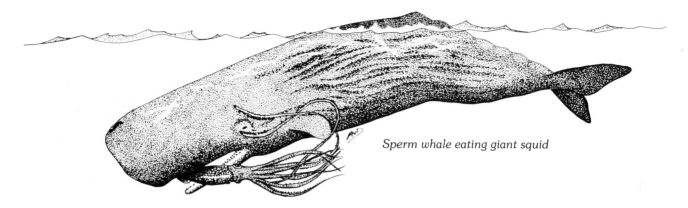

Sperm whale eating giant squid

niles. The older males tend to be solitary or in small groups, except during the breeding season.

A strange phenomenon is the periodic deliberate strandings on beaches of live toothed whales, and sometimes even baleen whales. Such animals, often in groups, become trapped in shallow water and are unable without assistance to return to the open ocean. Nowadays when reports of such strandings are received, considerable efforts are made to refloat live animals. Species to have stranded in groups in southern Africa include false killer whale and Risso's dolphin. Reasons for the strandings are unknown. There may be some social influence, because refloated animals sometimes return to other still-stranded members of their herd. In Florida, USA, some false killer whales once remained with an injured male for three days until he died, and only then did they disperse. Some scientists have advanced the theory that disturbance in the Earth's geomagnetic field, which these species use for navigation, causes the animals to become disorientated and may result in some strandings.

Baleen whales are characterized by the presence of baleen, or whalebone, in the mouth in the place of teeth. They also have two external blowholes, in contrast to the one of the toothed whales. The series of baleen plates embedded in the roof of the mouth serves to sieve plankton or fish from the water. The food of baleen whales ranges from krill and other zooplankton to shoaling fish. Toothed whales feed primarily on fish or cephalopods.

Baleen whales include the right whales, pygmy right whales and and rorquals, all of which are found off southern Africa, and the well-known gray whale of the North Pacific Ocean. The right whales have no dorsal fin or throat grooves, the pygmy rights a dorsal fin but no throat grooves, and the rorquals both a dorsal fin and throat grooves.

The right whale is distinguished by its lack of a dorsal fin, its strongly arched upper jaw and strongly bowed lower jaw, and the presence of callosities on the head. The callosities can be seen clearly on p. 91. Right whales migrate extensively, resulting in the species spreading towards the equator in winter and away from it, towards the poles, in summer. They are often found in shallow coastal waters. Cow and calf pairs are found in many southern African bays south of 32°S in spring, for example in Walker Bay – whale watchers at Hermanus often have a field day! In former years, their distribution extended farther north, to 20°S.

The humpback whale, a rorqual, is also a whale that often ventures inshore. It has very long, sometimes white or partly white, flippers, as much as a third of the body length, and a head that, in front of the blowholes, is flat and covered with knobs. The dorsal fin is often humped. These features can be seen on p. 94. The humpback whale breeds in tropical waters and has a more distinct pattern of migration than the right whale, passing southern Africa between its northern breeding season, which peaks about the middle of July, and southern summer feeding grounds.

Bryde's whale is interesting because it is the only baleen whale of the southern hemisphere that does not migrate to the Antarctic. One of its identifying features is its series of three prominent ridges running from near the blowholes to the snout (see p. 94). Bryde's whales are found over the southern African

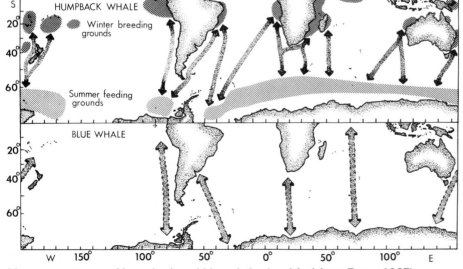

Migration patterns of humpback and blue whales (modified from Evans 1987)

Dusky dolphins

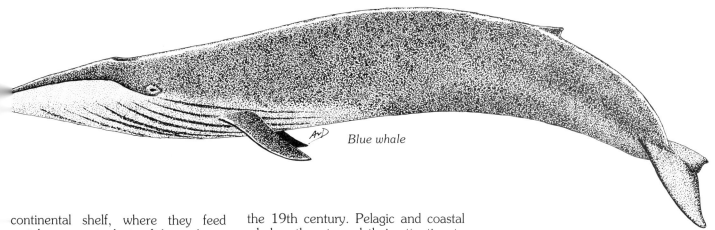
Blue whale

continental shelf, where they feed mainly on epipelagic fish such as anchovy, and farther offshore, where they eat euphausiids and mesopelagic fish. Individuals in these two regions are thought to belong to relatively discrete stocks.

Other species of baleen whale found in southern African waters include the blue, fin, sei and minke whales. All undertake extensive migrations between cold polar waters and warmer regions. The blue whale is the largest living animal, growing to more than 30 m and attaining weights in excess of 160 tons. It feeds almost exclusively on krill, and may eat up to 10 tons of this crustacean in a day. The fin whale is also enormous, growing to 27 m. Sei whales can attain 21 m and minke whales $10\frac{1}{2}$ m. Fin and sei whales are among the faster-swimming large whales and can achieve speeds in excess of 30 km/h over short distances.

North American and British whalers began operating off southern Africa towards the end of the 18th century. Although there was some limited shore-whaling at this time, this activity expanded considerably after the second British occupation of the Cape in 1806. Initially it was conducted from open boats, and the main target was the southern right whale. This species was approachable, relatively docile, and its buoyancy was such that it generally floated when dead. It had a high yield of oil, which was valuable as a lubricant and for lighting, and of whalebone. In every sense, therefore, it was the "right" whale. The whalebone was exported to England, where it was used in the manufacture of such objects as chair springs, hairbrush bristles and shoe horns. The southern right whale was also hunted in the open seas by pelagic whalers. Such exploitation over its entire range led to decreasing coastal catches of southern right whales during the 19th century. Pelagic and coastal whalers then turned their attention to humpback whales.

The humpback whale was both less valuable and harder to catch than the right whale. Also, harpooned animals often sank. Similar difficulties in the northern hemisphere led to the introduction of harpoon cannons fired from steam-driven catcher boats capable of winching dead whales to the surface. The harpoons had explosive heads, so whales could be crippled with the first shot. Such innovations led to the so-called modern era of whaling, and by 1929 there was no longer any whaling from open boats off southern Africa. Modern whalers operated from a number of localities around the subcontinent, including five in Angola, four in Moçambique, at Walvis Bay and Lüderitz on the Namib coast, and at Saldanha Bay, Cape Hangklip, Mossel Bay, Plettenberg Bay and Durban in South Africa. These involved both nearshore floating factories and land stations. During 1913, over 10 000 whales, many of which were humpbacks, were killed off southern Africa alone. Because the same species was also being exploited intensively elsewhere, it was not long before coastal catches plummeted.

Apart from the southern right and humpback whales, the only whales coming close to shore were those belonging to the inshore stock of Bryde's whale. This meant that, after the collapse of the two first-mentioned species, boats were forced to search farther out to sea. Landed catches of different whale species made off Namibia, off the Cape Province and off Natal from 1907 until 1975 are illustrated on p. 93. Off the Cape Province, blue whales and fin whales dominated catches from 1914 to 1930. Off Namibia there were good catches of blue whales, and off Natal it was mainly fin and sperm whales.

By the mid 1920s whaling boats had started to be constructed with stern slipways, allowing whales to be winched aboard the deck (see p. 7). This obviated the need for calm seas in which to flense whales alongside, and heralded the era of "ice-whaling". Large catches were soon being made in the Antarctic, and it was not long before the numbers of first blue whales (1930s), then fin whales (1950s), and in the 1960s sei whales decreased drastically. These patterns are reflected in the annual catches made off southern Africa.

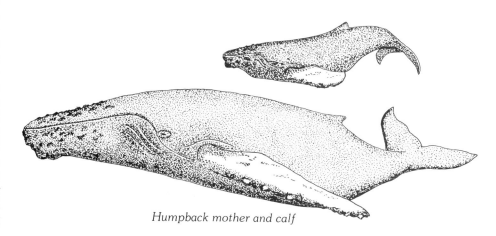
Humpback mother and calf

Indo-Pacific humpback dolphin **1**, Risso's dolphin **2** and southern right whale dolphin **3** may all be seen off southern Africa. Toothed whales that occur include sperm **4-5** and southern bottlenose **6** whales. The sperm whale has a huge box-like head and a single blowhole at the left of the head near the front. The blow projects obliquely forward. The southern bottlenose whale has a bulbous forehead and a bottle-nosed beak. It is rarely observed and there are few photographs of it

Baleen whales off southern Africa include southern right whales **1-2**, which have no dorsal fin, strongly bowed lower jaws and callosities on the head. The largest whale is the blue whale **3**. When viewed from above its head is U-shaped. Its huge body is a mottled, bluish grey. A subspecies recorded off southern Africa is the pygmy blue whale **4**. Fin **5** and sei **6** whales may also be encountered

Catches of sperm whales were then intensified, and smaller whales, such as minke and even killer whales, were sought in an attempt to maintain catches. However, profits were too small. Ultimately, South Africa ceased whaling at the end of 1975 after 184 virtually uninterrupted years. By now too, most other nations have also ceased whaling, and populations of some whale species have in fact started to recover.

Catches of whales off southern Africa, 1908-1975

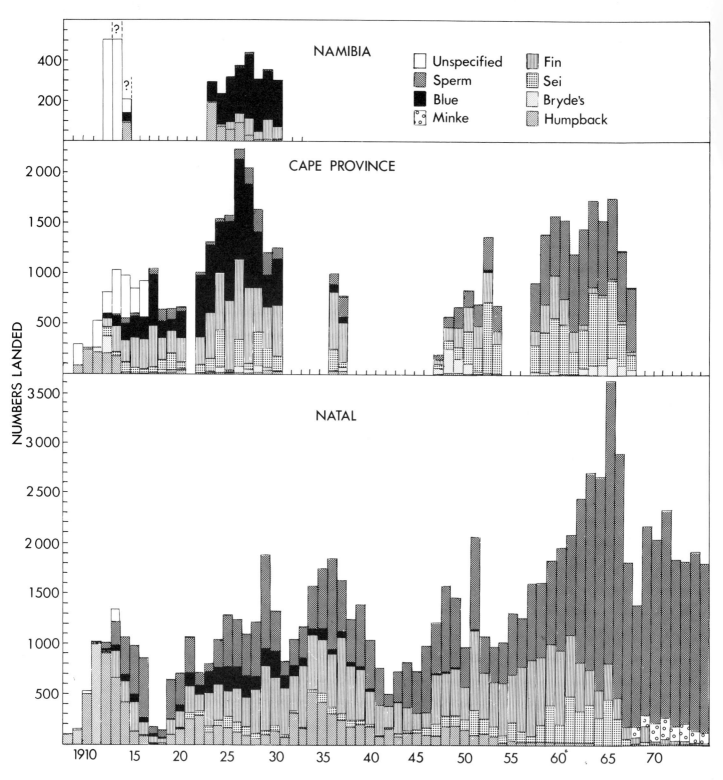

Two species of dolphin frequently seen off southern Africa's west coast are the dusky (top) and the Heaviside's (bottom). The latter is endemic to the region (photos B. M. Dyer)

Views from above of the Bryde's whale **1**, showing the characteristic ridges on its head, and of the minke whale **2-3**. Both species are found in southern African waters. Minke whales have a vertical blow **4**, characteristic of all rorquals. Humpback whales **5-7** have long flippers, often partly white, a dorsal fin that is humped and a head that, in front of the blowholes, is flat and covered with knobs

There is an important marine reserve at De Hoop **1**. South Africa's first marine national park was the Tsitsikamma Coastal National Park, a refuge for the Cape clawless otter **2** and a location of spectacular underwater seascapes **3**. It adjoins Nature's Valley **4**. The West Coast National Park includes the Langebaan Lagoon **5** and Malgas Island **6**. Sanctuaries in Maputaland protect the nesting sites of marine turtles. A loggerhead turtle moves towards the sea with Bhanga Nek in the background **7**

MARINE RESERVES

African black oystercatcher

Southern Africa has a large number of legally protected or semi-protected marine areas. In many instances portions of the adjoining mainland are also protected, thereby allowing for a continuation of natural processes across the land-sea interface.

Among the better-known marine reserves are two national parks, the Tsitsikamma Coastal and the West Coast National Parks. The Tsitsikamma Park was established in 1964 and was Africa's first coastal/marine national park. Its shoreline, between Plettenberg Bay and Humansdorp, is mainly a steeply rising rocky coastline bisected by a number of rivers flowing down often spectacular gorges. The mainland vegetation is a mixture of fynbos and evergreen forest. One of the more interesting mammals to find refuge at Tsitsikamma is the Cape clawless otter. This species lent its name to the Otter Trail, a hiking trail that winds along the coastline and on which otters are sometimes encountered. At Tsitsikamma, the Cape clawless otter makes its dens in thick scrub or holes along the coastline, homes that are referred to as holts. Most of its food comes from the sea, where fish and crabs are caught with remarkable agility. This makes it a good example of species that regularly cross the boundary between land and sea. The Cape clawless otter is equally at home in freshwater, and is also found in inland aquatic habitats.

The West Coast National Park includes part of the Langebaan Lagoon and some surrounding land, as well as five islands that lie within Saldanha Bay or the lagoon. These islands are important breeding localities for seabirds, which obtain their food and often also their nesting material from the sea. In addition to seabirds, the islands and the coastlines of Langebaan Lagoon and other estuaries are used by a large number of birds that feed at the high tide mark or in the intertidal region when the tide is out. Many of these birds are known as waders and they include both locally breeding species and birds that migrate to southern Africa from elsewhere. An example of the former is the African black oystercatcher, which nests at many of the islands and on some parts of the mainland and feeds on mussels and limpets obtained from the rocky shores. Some of the birds that travel to southern Africa do so from breeding grounds as far away as Siberia. Processes in that remote northern ecosystem are sometimes reflected in southern Africa. The curlew sandpiper nests on the ground in Siberia, where its eggs and chicks are vulnerable to the depredations of Arctic foxes. However, when lemmings are abundant in Siberia, the foxes have an ample supply of food and little need to plunder the sandpiper nests. Many of the chicks survive, and this is reflected in many young sandpipers visiting Langebaan Lagoon following Arctic breeding seasons in which lemmings have been abundant.

In addition to the five islands that form part of the West Coast National Park, 30 islands off the coasts of Namibia and South Africa are nature reserves administered by the Cape Nature Conservation. A portion of Robben Island is also run as a nature reserve and visitors can see African penguins that have recently recolonized the island.

The Cape Province has a number of marine reserves, among them the Cape of Good Hope Marine Reserve and the De Hoop Marine Reserve. Adjacent to both of these, substantial areas of mainland are also administered as reserves. The De Hoop Marine Reserve was established at the start of 1986, and there has already been a marked recovery in the population of South Africa's national fish, the galjoen, within its boundaries.

In Natal, Lake St Lucia and the St Lucia and Maputaland marine reserves, which protect much of Natal's northern coastline, are important conservation areas. Many aquatic birds breed at Lake St Lucia. Also, some beaches adjoining the open sea are used as nesting sites by loggerhead and leatherback turtles, the latter hauling its huge bulk of 700 kg up the beach to lay its eggs. Beneath the water, the marine reserves protect a variety of fish. Predators such as rockcod, potato bass and snapper kob have increased in abundance since the reserves were established. Also, the massive, but harmless, whale shark, which is the largest living fish and can attain 18 m, is regularly encountered. Diving in the marine reserves has become extremely popular.

Apart from the national parks and provincial reserves, there are areas that have been set aside for the protection of particular species, such as various rock-lobster sanctuaries. In 1989, the total length of shoreline off South Africa that afforded some measure of protection to marine life was over 600 km, slightly more than 20% of the country's total coastline.

The primary objective of the marine reserves is conservation. Conservation has been defined by the International Union for the Conservation of Nature and Natural Resources as "the management of human use of the biosphere so that it may yield the greatest sustained benefit to the present generation while maintaining its potential to meet the needs and aspirations of future generations". As well as protecting various species and habitats for future South Africans, the reserves have considerable potential for utilization by present-day inhabitants of the country. They provide recreational opportunities, such as enjoyment of underwater seascapes similar to those on p. 98. They are also useful for education, and of value for science. For example, unexploited populations provide information on the natural death rate of animals, which is needed to quantify the death rate caused by fishing in regions where the same animals are exploited.

It is necessary for the size of marine reserves to be sufficient to accomplish the objectives for which they were intended. For example, it has been shown that some reserves are too small to afford adequate protection to such fish species as dageraad and Roman, whereas reserves the size of the Tsitsikamma Coastal National Park make a meaningful contribution towards the conservation of the same species.

A second important requirement is that marine reserves should adequately represent the diversity of marine habitats and species found off southern Africa. Although a well-established network of marine reserves is already in place in the subcontinent, there is considerable scope for improvement. In particular, careful attention must be given to providing sufficient reserves to support southern Africa's burgeoning human population in future years.

Cape clawless otter

Pink ghost crab

A pygmy sperm whale makes a spectacular underwater sight **1**. Marine reserves provide opportunities for man to view the underwater world, and in the Tsitsikamma Coastal National Park, a Roman with a feather star **2** and groups of fish against attractive backgrounds **3-4** exemplify this statement. Crystal clarity of the water is evident in an underwater shot of the St Lucia Marine Reserve, in which goldies are swimming around plate and staghorn corals **5**. The striking coloration of organisms on the sea bed is shown by a sea fan and encrusting soft coral **6** and a cone shell **7**, both photographed in the De Hoop Marine Reserve

Species in the southern African marine system do not occur in isolation, but interact with other species and are influenced by man's activities. Dumping of fish off Namibia provided a ready source of food for Cape gannets in the 1960s **1**, and gannets have also scavenged behind demersal trawlers **2**. Killer whales can often be sighted near Sinclair and Mercury islands where they feed on seals and birds, such as the bank cormorant **3**. Seals consequently watch their

movements with keen interest **4**. Terrestrial predators, especially blackbacked jackals **5**, brown hyaenas **6** and lions **7** utilize seal pups and washed-up cetaceans as food, but prevent seabirds from breeding on the mainland to any large extent

99

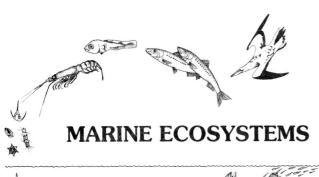

MARINE ECOSYSTEMS

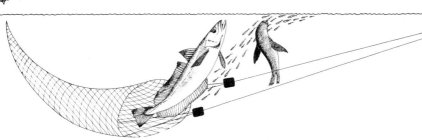

Many different forms of plant and animal life are found in southern Africa's seas. As on land there are the hunters and the hunted, and both have existed side by side for hundreds of years. Man is a relative newcomer to the marine ecosystems, and yet he is now exerting a considerable impact on them. For example, in recent decades modern fishing fleets have taken annual catches of 2-3 million tons off south-western Africa. How is this new and highly efficient predator influencing the functioning of the ecosystems?

The removal of vast quantities of fish must decrease the food available to other top predators, and it is thought that this has played a role in the decrease of some seabird populations. For example the number of African penguins has decreased drastically this century (see the Figure). Initially the decrease can be attributed to excessive collections of penguin eggs. However, the last collection of eggs was in 1967 but numbers of penguins have continued their downward trend. As discussed earlier, the recent decreases are likely to have resulted in part from a greatly reduced availability of pilchard, which was an abundant species and a major component of the African penguin's food in the 1950s.

In marked contrast to the penguin, numbers of South African fur seals have been increasing this century (see the Figure), despite their commercial exploitation. Seals have been able to adapt to decreased abundances of prey such as pilchard by feeding on other species, for example horse mackerel and bearded goby. They are able to dive to great depths and have a broader-based diet than penguins. The two trends may not be totally unrelated. Seals are now estimated to consume more than a million tons of food each year. Coupled with man's high harvest, this means that possibly three million tons less food is available to other predators than at the start of the century. Seals also occasionally eat penguins, and as their numbers have increased they have crowded onto some islands and displaced penguins from their breeding sites. Because man

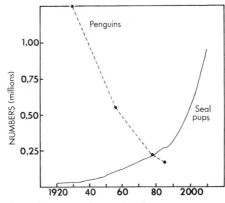

Trends in numbers of African penguins and South African fur seal pups in the 20th century

earlier cleared guano from many islands, penguins at those islands are now unable to construct the burrows which would have reduced this competition for space. On bare rock the penguins are no match for the much larger seals.

Man may have deprived some species of food but, through his wasteful activities, he has made food available to others. For example, killing seals for their skins and discarding the carcasses has provided food for scavengers such as giant petrels (see p. 71). Also, fishing operations bring fish to the surface and, when fish spill or are discarded, there is food for opportunistic seabirds and seals (see pp. 26, 30 and 99). This too would have advantaged the South African fur seal over the African penguin.

If a species is fished to a low level of abundance, the food that species utilized becomes available to other organisms in the ecosystem and they in turn may increase in abundance. Species such as anchovy, horse mackerel and bearded goby probably increased following the collapse of the pilchard, and this is likely to have favoured predators such as bank cormorants and seals.

Clearly man has the potential to influence the marine ecosystems in a number of ways, although the manner in which he is doing so is still poorly understood. However, it would be most unfortunate if trends such as the downward slide of the African penguin were allowed to continue. Despite the lack of knowledge, some form of intervention in the functioning of the ecosystems is called for and has been implemented. Thus, seals displacing rare or endangered seabirds from breeding sites are now disturbed so that they will move elsewhere to pup.

THE HERITAGE

Cape clawless otter eating fish

A small part of the rich marine heritage that belongs to southern Africa, and which has been the theme of this book, is shown on p. 102. Large resources of fish and shellfish support commercial fisheries, supply much-needed protein – often in the form of top-quality dishes – earn substantial foreign revenue, and provide many people with work. In South Africa (excluding Walvis Bay), the commercial fishing industry employs some 26 000 people, about half of this number at sea. Most of the remainder work in plants where the fish is offloaded and processed, of which there are more than 100. Additional to these jobs, a considerable infrastructure is required to support the industry, for example through the construction and maintenance of boats and machinery. South Africa has more than 6 000 fishing boats of various sizes, and the replacement value of the fleet is over R3 000 million.

Commercial fishing is but one aspect of the utilization of marine resources; for vast numbers of people the seas provide endless hours of recreational enjoyment. One of the most popular forms of marine recreation is angling, whether from the shore or from a boat. There are nearly half a million anglers in South Africa. Presumably it is the thrill of catching the evening's meal that attracts them, and there are many fish and shellfish that suit that bill. In the 1988/89 season almost 40 000 licences were issued to the public to catch rock lobster, and more than 20 000 to dive for abalone. South Africa has about 4 000 spearfishermen. Divers can enjoy spectacular vistas underwater, and there is always the possibility of encountering something unusual and exciting – such as the otter shown on this page. Again there is a sizeable infrastructure, for example those who supply tackle or holiday accommodation.

For many people enjoyment comes from observing, and perhaps photographing, the numerous species of seabirds and marine mammals that are to be seen. Perhaps nothing enlivens the seascape more than a group of dolphins or one of the large whales. At islands and some mainland sites, the antics of seabirds and seals at their breeding localities may be watched. Lambert's Bay is a particularly good locality for viewing seabirds at close proximity, and it is to be hoped that access to other of southern Africa's islands will soon also be allowed. Breeding birds are sensitive to disturbance, and great care must be taken to avoid interference with natural processes. In Natal it is possible to observe sea turtles as they come ashore to nest.

The seas and their resources provide opportunities for education, for scientific and historical study, for some cultural activities to continue and perhaps for new ones to develop. If treasures of the past are to be preserved, now is the time to act. For example, although commercial exploitation of African penguin eggs terminated as recently as 1967, after intensive investigations only one of the tools used for collecting the eggs has so far been traced – that photographed on this page. Similarly, very few of the instruments formerly used at islands for rendering seal blubber to oil remain.

The marine heritage is a gift to be appreciated; one to be used with wisdom, treasured, enjoyed and handed on to succeeding generations. Fortunately, this has been recognized by the country's legislators. Some acts provide for the conservation of specific resources, for example seabirds and seals. The Sea Fishery Act provides guidelines for general policy, specifically the conservation of the country's marine ecosystems and the sustained utilization of their resources "to the greatest benefit of the present and future inhabitants of the Republic". Attaining these objectives is a challenge for scientists, fishermen, industrialists, tourists, law enforcement officers – indeed for all who are involved to a greater or lesser extent with the splendour of southern Africa's seas.

Instruments that were formerly used for collecting and packaging penguin eggs

Southern Africa's marine resources are a rich, natural heritage. It is to be hoped that marine life off southern Africa will continue to be in bountiful supply: an abundance of spiny lobsters **1**, a purse-seiner laden with anchovy **2**, a pleasing variety in the catch of a Natal linefisherman **3**, a common dolphin **4**, a view of endemic Cape gannets, African penguins and South African fur seals at Mercury Island in 1985 **5**, and a southern right whale **6**

INDEX

Snoek strike

Page numbers in bold indicate a photograph or an illustration

Abalone **20**, **22**, 101
Albatross, Blackbrowed **70**, 72, **73**
　Darkmantled sooty **71**
　Greyheaded **71**
　Lightmantled sooty **71**
　Shy **70**, 72, 73
　Wandering **70**, **73**
　Yellownosed **70**, 72
Amphipod **13**
Anchovy 9, **17**, 24, 25, **27**, 37, 61, 65, 89, 100, 102
Angler, angling **7**, 8, 45, 48, 101
Anglerfish – see Monkfish

Bass, Brindle **39**
　Potato **39**, 96
Billfish 37
Blacktail **38**
Booby 60, 72, 73
Bream, Bronze 32
　River **38**
Butterfish **38**

Cat, Feral 57
Chaetognath 13
Chimaera 41
Chub mackerel 25
Coachman, Schooling **39**
Colonial breeding 68, 69
Copepod 13, **15**
Cormorant, Bank 60, 61, **63**, **64**, 68, **99**, 100
　Cape **60**, 61, **63**, 64, 65, **67**, 68
　Crowned **6**, **7**, **10**, 60, **63**, 64, **66**, 68
　Whitebreasted 60, 64, **66**
Cumacea 13
Crab **13**, **15**, 49, **54**, 77, 96, **97**
Current, Agulhas 9, 11, 85
　Benguela 9, 11, 84
　Moçambique 9
Cuttlefish, **22**, 85
　Common **20**

Dageraad 32, **38**, **40**, 97
Darter 60
Diatom 12
Dinoflagellate **12**
Dogfish – see Shark

Dolphin, 45, 101
　Bottlenose 81, 84, 85, **87**
　Common 81, 84, 85, **87**, **102**
　Dusky 81, 84, 85, **86**, **88**, **93**
　False killer whale 81, 84, **86**, 88
　Fraser's 81, 85
　Heaviside's 81, **83**, 84, 85, **93**
　Indo-Pacific humpback 81, 85, **90**
　Killer whale 81, **84**, 85, **86**, 92, **99**
　Long-finned pilot whale 81, 84, **86**
　Melon-headed whale 81
　Pygmy killer whale 81
　Risso's 81, 84, 85, 88, **90**
　Rough-toothed 81
　Short-finned pilot whale 81
　Southern right whale 81, **84**, 85, **90**
　Spinner 81, 85, **87**
　Spotted 81, 85, **87**
　Striped 81, 85, **87**
Dragonet, Ladder **36**

Egret **10**, 68
Elf **40**
Estuary **8**, **10**
Euphausiid **13**, **15**, 89

Fishery, 6, 9, 101
　Abalone 20
　Beach-seine 21, 32, **34**
　Bottom trawl 17, **19**, **28**, 29, 30, 73
　Line 32, 33
　Lobster **17**, **23**
　Longline **19**, 29, **31**
　Midwater trawl 17, **19**, 28, 73
　Prawn 17, 21
　Purse-seine 17, **19**, **24**, **26**, 73, 85, **102**
　Squid 17, 20
Fishing, Artisanal 48, **50**
　Gear 6, **17**, 21, **24**, **28**,
　Recreational 32, 48, 50, 101
　Regulations 17, 48
　Subsistence 48, 50
Fransmadam **42**
Frigate bird 52, 60, 72, 73
Fulmar, Antarctic 72, **74**

Galjoen 32, 48, 96

Gannet, Australasian 60, **72**, **73**
　Cape **5**, **16**, 60, 61, **62**, **64**, 68, 72, **73**, **80**, **99**, **102**
Garrick **33**
Geelbek 33, 40
Genet, Large-spotted **54**
Goby 61, 64, 100
Grebe, Blacknecked 72, 73
Guano 53, 60, 61, 62, 64, 65, 100
Gull, 72
　Greyheaded 65, **67**, **68**
　Hartlaub's 65, **66**, **67**, 68
　Kelp 30, **61**, 65, **66**, 68, **69**, 73
　Sabine's **75**

Hake 17, 21, 28, 29, **30**, 37, 40, 73
Heron **10**, 68
Horse mackerel **17**, 25, **27**, 28, **30**, 61, 100
Hottentot **19**, 32, **33**, **34**, 40
Hyaena, Brown 53, 77, **99**

Ibis, Sacred 68
Island, Coastal 53, **55**, 64, 68, 96, 99, 101, 102

Jacopever 31
Jackal, Blackbacked 53, 65, 77, **79**, **99**
Jellyfish 13, **15**, 26, 49
John Brown **35**

Kingfish **42**
Kingklip **17**, 28, **29**, **30**, **31**, 37
Kob **33**, **40**, 96

Lanternfish **36**, 84
Leervis – see Garrick
Leopard 53
Linefish 32, 33, 37, 40, 102
Lion 53, 77, **99**
Lobster, Spiny 17, **19**, **21**, **23**, 28, 77, 96, 101, **102**

Mermaid's purse **44**
Mollusc 13, 20
Monkfish **17**, 28, **30**, **36**, 37
Mullet 22, 32, 85
Musselcracker **35**
Mysid **13**

Nautilus 20, **22**
Net, 80
　Beach-seine 32, **34**
　Bottom trawl 17, **28**, 29, 73
　Hoop 21
　Midwater trawl 17, 73
　Purse-seine 17, **24**, 73, 85
　Shark – see Shark net

Octopus, 21, 85
　Giant **20**
Onderbaadjie – see Lanternfish
Otter, Cape clawless **95**, 96, **97**, **101**

103

Oystercatcher, African black **96**

Pelican, White **10**, **60**, 64, 65, **66**
Penguin, 53, 72, 73,
    African **18**, 55, **56**, **57**, **58**, **59**, 61, 64, 68, **80**, 96, 100, 101, **102**
        Jackass, see African
        King **59**
        Macaroni **59**
        Rockhopper **59**
Perlemoen – see Abalone
Petrel, Antarctic **74**
    Greatwinged 72
    Grey **74**
    Northern giant **71**, 73, 100
    Pintado **72**, 73, **74**
    Softplumaged **74**
    Southern giant **71**, 73, 100
    Whitechinned 72, 73, **74**
Phalarope 72, 73
Photosynthesis 9, 12
Pilchard **17**, 24, **25**, 26, **27**, 37, 57, 61, 64, 84, 100
Plankton, 9
    Phyto 12, 13,
    Zoo 12, 13, 15, 88,
Platform, Guano 53, 64, 65
Poenskop 32, **37**, 40
Pollution 8, 16, **18**, 57
Prawns **8**, 17, **21**, 77
Prion, 72, 73
    Broadbilled **74**
    Fairy **74**

Ray 44
Recreation 6, 32, 33, 37, 40, 48, 97, 101
Red-eye – see Round herring
Red water 12, **14**
Reserves, Marine **95**, 96, 97, **98**
Rockcod 40, **42**, 96
Roman 32, 33, **34**, **38**, **40**, 97, **98**
Round herring 25

Sailfish **33**
Sandpiper, Curlew 96
Santer **48**
Sardine – see Pilchard
Sardine run 25, **27**, 84
Seabream 40
Seahorse, Knysna 36, **37**
Seal, 44, 85
    Crabeater **76**
    Leopard 76, **79**
    South African fur 6, **7**, **30**, 46, 53, 61, 68, 73, 76, **77**, **78**, **79**, 80, **82**, **99**, 100, 101, **102**
    Southern elephant 76, **79**
    Subantarctic fur 76, **79**
Sealing 73, 80, **82**
Shark, 21,
    Blue **43**
    Blunthead **47**
    Dusky 45, **47**
    Great white **41**, 44, 45, **46**
    Izak catshark **43**
    Ragged-tooth 44, **46**
    Scalloped hammerhead **47**
    Shortnose spiny dogfish **43**, **45**
    Spotted gully **43**
    St Joseph **41**
    Tiger **44**, 45
    Whale **41**, 44, 96
    Zambezi 41, 45
Shark net 45, 84, 85
Shearwater, 73
    Cory's **75**
    Great **72**
    Sooty 72, **75**
Sheathbill, American 72
Shrimp, Mantis **15**
Skate,
    Biscuit **44**
    Spearnose **45**
Skua, 72
    Pomarine **73**, **75**
    South Polar **75**
    Subantarctic **73**, **75**
Slinger 32, **35**
Snoek 32, 33, 37, **80**, **103**
Sole 17, 28
Squid, 17, 85, **88**
    Chokka 20, **22**
Steenbras, Red **38**, **48**
Stingray, **43**
    Honeycomb **41**
Storm petrel, 72
    Blackbellied 73, **75**
    Wilson's **75**
Strepie **39**
Stumpnose, Red **32**, **34**, **50**
Sunfish **36**

Tern, Antarctic **75**
    Arctic 73, **75**
    Caspian 65, **67**
    Common **72**
    Damara 65, **67**, 68
    Roseate 65, **67**
    Sandwich 72
    Swift **65**, **67**, 68, **69**
Trawl – see Net
Tropic bird 60, 72, 73
Tuna, 32, 33, 37, 85
    Bigeye **19**
    Skipjack **42**
    Longfin **29**, **31**
    Yellowfin **40**
Tunicate 13
Turtle, 45
    Green **49**, **51**, **52**, **54**
    Hawksbill 49
    Leatherback 49, **51**, 52, **54**, 96
    Loggerhead **49**, **51**, 52, **95**, 96
    Olive Ridley 49

Upwelling **9**, 12, 32, 84

Wahoo **42**
Whale, 6, 13, 73, **99**, 101
    Arnoux's beaked 81
    Blainville's beaked 81
    Blue 81, 88, **89**, **91**, 92
    Bryde's 81, 88, 89, 92, **94**
    Cuvier's beaked 81
    Dwarf sperm 81, 85
    Fin 81, 89, **91**, 92
    Gray's beaked 81
    Hector's beaked 81
    Humpback 81, 86, 88, **89**, 92, **94**
    Killer – see Dolphin
    Layard's beaked **81**
    Minke **7**, 81, 89, 92, **94**
    Pilot – see Dolphin
    Pygmy blue **91**
    Pygmy right 81, 88
    Pygmy sperm 81, 85, **98**
    Sei 81, 89, **91**, 92
    Southern bottlenose 81, **90**
    Southern right 81, 88, 89, **91**, **102**
    Sperm 81, 85, **88**, **90**, 92
    True's beaked 81
Whaling 7, 89, 92
Whalefish **39**

Yellowtail 33, **35**, **40**

Zebra **35**

© Copyright 1992

Vlaeberg Publishers, P. O Box 15034 Vlaeberg, 8018, South Africa

All rights reserved. No part of this publication may be copied and/or published by means of photocopying, microfilm, or any other means without the consent of the publishers.

Payne, A I L and Crawford, R J M
1992 Secrets of the Seas

Production and design:
Wim Reinders & Associates, Cape Town
Cover design: Abdul Amien
Typeset: Hirt and Carter, Cape Town
ISBN 0-947461-14-0